ICT for Transport

NECTAR SERIES ON TRANSPORTATION AND COMMUNICATIONS NETWORKS RESEARCH

Series Editor: Aura Reggiani, *Professor of Economic Policy, University of Bologna, Italy*

NECTAR (Network on European Communications and Transport Activities Research) is an international scientific, interdisciplinary association with a network culture. Its primary objective is to foster research collaboration and the exchange of information between experts in the fields of transport, communication and mobility.

NECTAR members study the behaviour of individuals, groups and governments within a spatial framework. They bring a wide variety of perspectives to analyse the challenges facing transport and communication, and the impact these challenges have on society at all levels of spatial aggregation.

This series acts as a companion to, and an expansion of, activities of NECTAR. The volumes in the series are broad in their scope with the intention of disseminating some of the work of the association. The contributions come from all parts of the world and the range of topics covered is extensive, reflecting the breadth and continuously changing nature of issues that confront researchers and practitioners involved in spatial and transport analysis.

Titles in the series include:

Transportation and Economic Development Challenges
Edited by Kenneth Button and Aura Reggiani

City Distribution and Urban Freight Transport
Multiple Perspectives
Edited by Cathy Macharis and Sandra Melo

Accessibility Analysis and Transport Planning
Challenges for Europe and North America
Edited by Karst T. Geurs, Kevin J. Krizek and Aura Reggiani

Smart Transport Networks
Market Structure, Sustainability and Decision Making
Edited by Thomas Vanoutrive and Ann Verhetsel

Accessibility and Spatial Interaction
Edited by Ana Condeço-Melhorado, Aura Reggiani and Javier Gutiérrez Puebla

ICT for Transport
Opportunities and Threats
Edited by Nikolas Thomopoulos, Moshe Givoni and Piet Rietveld

ICT for Transport

Opportunities and Threats

Edited by

Nikolas Thomopoulos

LSE Cities, London School of Economics and Political Science, UK

Moshe Givoni

Director, Transport Research Unit, Department of Geography and the Human Environment, Tel-Aviv University, Israel

Piet Rietveld

former Professor of Transport Economics, VU University Amsterdam, the Netherlands

NECTAR SERIES ON TRANSPORTATION AND COMMUNICATIONS NETWORKS RESEARCH

EE Edward Elgar PUBLISHING

Cheltenham, UK • Northampton, MA, USA

Published by
Edward Elgar Publishing Limited
The Lypiatts
15 Lansdown Road
Cheltenham
Glos GL50 2JA
UK

Edward Elgar Publishing, Inc.
William Pratt House
9 Dewey Court
Northampton
Massachusetts 01060
USA

A catalogue record for this book
is available from the British Library

Library of Congress Control Number: 2014959466

This book is available electronically in the **Elgar**online
Economics subject collection
DOI 10.4337/9781783471294

MIX
Paper from
responsible sources
FSC® C013056
FSC
www.fsc.org

ISBN 978 1 78347 128 7 (cased)
ISBN 978 1 78347 129 4 (eBook)

Typeset by Servis Filmsetting Ltd, Stockport, Cheshire
Printed and bound in Great Britain by T.J. International Ltd, Padstow

to our families

to my father who showed me how to adapt and take up new challenges
living up to my values
Nikolas

Contents

Contributors

Margarita Angelidou is a researcher at the URENIO Research Unit of the Aristotle University of Thessaloniki, Greece. She has contributed to numerous EU and national research projects and scientific publications. She holds degrees in Architectural Engineering (BSc), Urban Planning and Development (MSc), Management of Urban Recourses (MSc) and Strategic Planning for the Development of Smart Cities (PhD). Her main research interests include urban and social development, urban technologies and innovation ecosystems.

Erel Avineri is Head of the Infrastructure Systems Engineering and Management Department at the Afeka – Tel-Aviv Academic College of Engineering, and the Head of ACITRAL – Afeka Centre for Infrastructure, Transportation and Logistics. He holds degrees in Industrial Engineering and Management (BSc) and Transportation Sciences (MSc, PhD) from Technion – Israel Institute of Technology. His main research areas are travel behaviour, road safety behaviour, intelligent transport systems, and transport policy.

Eran Ben-Elia is a Senior Lecturer, Department of Geography and Environmental Development at Ben-Gurion University of the Negev, Israel. He holds a PhD in Transportation Science from Technion – Israel Institute of Technology. His research interests lie in the interconnections between space, emerging technologies and human behaviours, in particular using ICT to understand urban mobility patterns and to encourage sustainable mobility choices.

Kuan Butts has a Master's in City Planning (MCP) degree from MIT's Department of Urban Studies and Planning, where he studied the relationship between mobile technologies and urban analyses, primarily in relation to transportation. Kuan has previous professional planning experience working primarily in midwestern American cities. While at MIT, under the Flocktracker project, Kuan led digital transit mapping efforts in Dhaka, Bangladesh and Mexico City, Mexico.

Arturo Cadena is an urban planning student at the Universidad Nacional Autónoma de México (UNAM) and has worked with local government and on research projects focusing on 'Urban Social Production'. He was a geospatial analyst on the Mexico City Flocktracker project and is a founding member of Urban Launchpad Mexico.

Cristian Camusso is a Postdoctoral Research Fellow at Politecnico di Torino, Interuniversity Department of Regional and Urban Studies and Planning. He received a degree in Civil Engineering-Transport and in 2009 a PhD in Environment and Territory at Politecnico di Torino, whilst in 2011 he was ECTRI representative of the same institution. He has worked on several European projects (COST355, COST356, SmartWay, Opticities, Alpinfonet) and his scientific interests focus mainly on transport and noise indicators, travel behaviour and the effects of Intelligent Transport Systems on users' behaviour.

Maria Chatziathanasiou is a Civil Engineer, holding a Master's in Design, Organization and Management of Transportation Systems. Since 2007 she has been working as a transport engineer, participating in a number of traffic studies and research projects. She is the author of several papers published in national and international conference proceedings and peer-reviewed journals.

Elizabeth Dodds is a consultant with the Open Aid Partnership (OAP), a multi-stakeholder initiative hosted in the World Bank Group's Innovation Labs. In her current position with the World Bank, she supports OAP efforts to engage civil society, journalists and citizens in development decision-making through the use of open data. Prior to joining the Bank, she gained private sector experience as a financial regulatory associate for BNY Mellon Asset Servicing. Elizabeth received a Master's degree in public administration from the London School of Economics and a Bachelor's degree in International Relations and French from Colgate University.

Pelle Envall is CEO of TUB (Trafikutredningsbyrån), a Stockholm-based consultancy focusing on transport planning and urban development. He has been the coordinator of the Swedish CyCity Research Programme (2009–14) as well as the principal developer of the BikeRoute analysis tool. Dr Envall holds a PhD in Transport Planning from the University of Leeds and during his career he has managed research grants of 1 million euros, analysing car parking requirements in new developments.

Moshe Givoni is Head of the Transport Research Unit (TRU) and a Senior Lecturer in the Department of Geography and Human Environment, Tel Aviv University. He is also a Visiting Research Associate at the Transport Studies Unit, University of Oxford and an Associate Editor for the journal *Transport Reviews*. He gained his PhD at the Bartlett School of Planning,

University College London and received a Marie Curie postdoctoral fellowship which was undertaken at the VU University, Amsterdam. His academic background includes degrees in Economics and Geography (BA) and Business Administration (MBA) from Tel Aviv University. He is co-editor of the book *Moving Towards Low Carbon Mobility*, together with David Banister (Edward Elgar, 2013).

Bernard Gyergyay is an Urban Mobility Consultant at Rupprecht Consult in Germany where he coordinates EU-funded sustainable urban mobility projects. Bernard worked as a transport planner in the Transport for London Behaviour Change Department where he managed projects that generate demand for the Cycle Superhighway. He has studied and researched transport policy and travel behaviour at the Institute for Transport Studies (University of Leeds) as well as at the Centre for Transport Studies (Imperial College London and University College London).

Lee D. Han is Professor in Civil Engineering at the University of Tennessee, Knoxville. He received his Doctorate from the University of California, Berkeley. He is also an Oak Ridge National Laboratory Collaborating Scientist. His fields of expertise include traffic engineering, application of advanced technologies to transportation and system modelling and simulation, among others.

Heidi Herzogenrath-Amelung is Senior Lecturer in Media & Communications Theory at the University of Westminster, and has recently completed her PhD on Martin Heidegger's philosophy of technology and digital surveillance. Her research focuses on critical theories of the media and ICTs, surveillance and philosophy of technology. She has co-founded and coordinated research networks on surveillance and critical approaches to ICTs.

Nicos Komninos is Professor of Urban Development and Innovation Policy at the Aristotle University of Thessaloniki and founder and director of URENIO Research; he has coordinated about ninety research projects on innovation ecosystems and intelligent cities. He has published extensively on urban and regional development, innovation territories, smart and intelligent cities, including the trilogy *Intelligent Cities: Innovation, Knowledge Systems and Digital Spaces* (2002), *Intelligent Cities and Globalisation of Innovation Networks* (2008), and *The Age of Intelligent Cities: Smart Environments and Innovation-for-all Strategies* (2014).

Anna Kortsari is a research associate at the Hellenic Institute of Transport at the Centre for Research and Technology Hellas. She holds a degree in Civil Engineering from the Aristotle University of Thessaloniki and a

Master's in Business Administration from the University of Macedonia in Greece. She is currently a PhD candidate focusing on the field of railway transport. She has participated in numerous research projects on railway transport systems, mobility management, seasonal transport demand management, and intelligent transport systems, while contributing to several scientific papers.

Xavier Leal holds a Master's in Aeronautical Management and a Master's in Project Management. He works as Research Development manager in the Telecommunications Department of Universitat Autònoma de Barcelona (UAB). He is specialized in innovation process analysis at firm, national and system levels and has participated as technical advisor in a number of projects funded by the European Commission. Currently he is the coordinator of the FP7 projects T-TRANS (www.ttransnetwork.eu) and INOCAST (www.inocast.eu).

Wei Lu is Postdoctoral Research Associate at Oak Ridge National Laboratory. He received his Doctorate from the University of Tennessee, Knoxville and his Master's from the University of Science and Technology of China. His research interests include connected vehicle technology, traffic simulation and modelling, sustainable transportation and intelligent transportation systems.

Daniel Palencia is a civil engineer currently spending most of his time programming and riding his bike around Mexico City. He is a programmer on the Flocktracker Android app project and has worked on several Urban Launchpad projects since 2013.

Isidoros A. Passas has a degree in Engineering on Computer Systems from the School of Engineering of the University of Sussex, and a postgraduate diploma in Business Administration from the University of Sheffield. His scientific interests involve the use of Information and Communication Technologies for the knowledge enhancement and competitiveness of SMEs. His core research fields are strategic intelligence, open knowledge diffusion, data mining and data visualization for regions and enterprises.

Cristina Pronello is currently Associate Professor at the Politecnico di Torino, Interuniversity Department of Regional and Urban Studies and Planning and was previously Professor at Laboratoire d'Economie des Transports (LET) Université Lumière Lyon 2. She has been the Chair of the COST Domain Committee Transport and Urban Development from 2006 to 2014, a member of the Scientific Advisory Board of the JPI Urban Europe and has also served on the ECTRI board from 2005 to 2013. Her scientific interests revolve around travel behaviour, the effects of Intelligent Transport Systems on users' behaviour, mobility surveys and the environmental impacts of transport systems.

Piet Rietveld (1952–2013) studied econometrics at Erasmus University, Rotterdam (cum laude degree) and received his PhD in economics at VU University Amsterdam. He worked at the International Institute of Applied Systems Analysis (Austria) and was the research coordinator at Universitas Kristen Satya Wacana in Salatiga, Indonesia. Since 1990 he was professor in Transport Economics at the Faculty of Economics and Business Administration, VU University Amsterdam. He was a fellow of the Tinbergen Institute, the Regional Science Association International (RSAI) and the Netherlands Institute for Transport Policy Analysis (KiM). Furthermore, he was a member of the Royal Netherlands Academy of Arts and Sciences (KNAW). Piet was Head of the Department of Spatial Economics at VU University Amsterdam between 2002 and 2013. As a researcher, he made groundbreaking contributions to the fields of spatial, transport and environmental economics, on a wide variety of themes including transport and regional development, valuation, transport pricing, public transport, transport and environment, land-use modelling, and policy assessment.

Maria S. Schoina is a graduate of the Department of Civil Engineering of Aristotle University of Thessaloniki (AUTh) and has been involved for several years in structural design studies. Since 2000, she has been a member of the Urenio Research Unit of AUTh, participating in the implementation and management of a number of EU and national research projects in the fields of regional development, innovation, knowledge management, business intelligence and green buildings.

Elena Sefertzi is Associate Professor of Spatial Systems and Innovation Strategies at the Department of Spatial Planning and Development Engineering, Aristotle University of Thessaloniki. She has participated as expert or coordinator in about fifty research projects of the URENIO research unit on innovation strategies and regional development, and she is the author of articles and books on regional innovation systems and strategies, strategic planning, technology transfer and industrial development.

Nikolas Thomopoulos is a researcher at LSE Cities. Previously he was a researcher at the Institute for Transport Studies at the University of Leeds where he was also a Marie Curie EST Fellow. He has a multidisciplinary background gained through studies in Greece, Germany and the UK and has contributed in a range of European FP-6 and FP-7 research projects, whilst having also lectured on economics, business, environmental policy and transport appraisal. In 2013–14 he was the Chair of a Regional Studies Association Research Network on 'smart' and 'green' Growth and he organized a range of ICT-related workshops. His research interests revolve around the socio-economic and business impacts of ICT for transport, the review of economic and environmental assessment methods and the evaluation of urban and regional policy.

Pinelopi Troullinou is a PhD researcher at the Open University Business School. Her research focuses on the possibilities of everyday resistance to surveillance practices occurring via the use of digital technologies. Her work is funded by the Social Sciences and Humanities Research Council of Canada in the context of The New Transparency: Surveillance and Social Sorting project. Previously she worked as a Research Assistant for the EU FP7 Project, ICT Ethics.

Yannis Tyrinopoulos is Assistant Professor at the Technological Educational Institute of Athens, Greece. He is a Mathematician–Transportation Engineer and his research record includes the participation in a large number of European R&D projects while he has also participated in many expert groups, committees and panels, such as the TRB AP020 Committee on Emerging and Innovative Public Transport and Technologies. He is the author of more than 60 publications, all of which have been published in peer-reviewed scientific journals or in international conference proceedings.

Qiyang Xu is a consultant in the World Bank Group's Innovation Labs. She has worked in the areas of Open Data and ICT and geospatial innovations in international development. She leads geospatial activities for various initiatives across multiple sectors and countries, including spatial and statistical analysis, data management and visualization, web-mapping and crowdsourcing applications. Between 2011 and 2012, she worked for the Foundation Center in New York, facilitating spatial analysis for CSOs and NGOs using open source applications. Qiyang received a Master's degree in Geography from the University of Maryland and a Bachelor's degree in Geographic Information Systems and German from Zhejiang University.

P. Christopher Zegras is Associate Professor of Transportation and Urban Planning in the Department of Urban Studies and Planning and the Engineering Systems Division at MIT. He has over two decades of experience working on urban mobility data collection, modelling, planning and policy analysis in the USA, Latin America, Europe and Asia. He leads the Mobility Futures Collaborative, through which the Flocktracker project was developed.

Foreword

It is both a great pleasure and a sad task to write a foreword for the current volume *ICT for Transport: Opportunities and Threats* edited by Nikolas Thomopoulos, Moshe Givoni and Piet Rietveld.

It is a pleasure because the volume provides a rich and comprehensive insight into the ever-increasing role of ICT (Information and Communication Technologies) in the evolving European and international space-economy, with a particular view on its impact on, and intersection with, the transport sector. A core element of this book is the association between ICT and the sustainable mobility paradigm, and thus the increasing link of ICT with emerging challenges such as the provision of incentives to positively influence travel behaviour, crowdsourced geolocation data accuracy or the digital divide and the respective challenges for policy makers. In an attempt to answer pressing research and policy questions, this book offers a valuable insight into the similarities and differences in the ICT needs of both developed and developing countries. In this context, a novel contribution of this volume is the impact of ICT on different transport modes, for example car, bus, walking and cycling, reviewed through empirical findings and the views of practitioners and scholars with an international background. This volume is an output of NECTAR (Network on European Communications and Transport Activities Research), which is a network of multidisciplinary researchers that brings together a wide range of perspectives on transportation, mobility and communication issues and their impacts on society. Thus this volume puts a truly international and holistic ICT view in a transport perspective, with contributions by authors from different parts of the world.

It is a sadness at the same time though since it is the last volume edited by Piet Rietveld (1952–2013). Therefore, this volume can be considered to honour his memory.

Piet Rietveld chaired NECTAR between 2002 and 2007, and it was through our common scientific interest in transport network evolution and the plethora of research and networking opportunities provided through NECTAR, that I had the opportunity and privilege to work closely with him over the last 12 years. During those years, Piet revealed an exceptional academic ability, which allowed him to diversify his research focus within the broad field of transport, from accessibility to equity, from climate change to infrastructure networks, from cycling to ICT and multimodal analyses. His contributions were diverse, characterised by a variety of approaches, namely theoretical, methodological, empirical or economic policy ones.

During 2008–13, Piet Rietveld was also the coordinator of the NECTAR Cluster 1 on Networks, together with Moshe Givoni and Bart Jourquin (for Europe) and Andrew Goetz (for the United States). Over these six years, Piet's Cluster was always extremely active, organising several meetings in Europe where new scientific ideas were promoted and expounded.

The idea about the present book was a result of Piet's initiative and was first discussed during one of these Cluster 1 meetings, since the link between ICT and transport was one of his favourites. Given this opportunity, I wish to thank Nikolas Thomopoulos and Moshe Givoni, who have conscientiously continued and fulfilled the scientific research path of this book, which was originally followed jointly with Piet, by producing an excellent landmark in this emerging field of ICT for transport.

This book can therefore be considered as the last legacy to us from a brilliant scientist who passed away all too soon.

Aura Reggiani
Professor of Economic Policy
University of Bologna, Italy
NECTAR Chair 2008–13
Editor of the NECTAR Series on
Transportation and Communication Networks Research

Preface

Since NECTAR combines research about both transport and communications, this book series was considered as an ideal setting to expand on the topic of ICT for transport. The idea about this book was first discussed at a NECTAR Cluster 1 Networks meeting between the editors. Piet Rietveld's role was instrumental, so a NECTAR Cluster 1 workshop took place, focusing on the intersection of transport with social issues. Interestingly, a lot of contributions in that workshop addressed the challenges of contemporary technological developments for the transport sector. Such concerns revolved around privacy or the values used within transport appraisal. However, one of the workshop conclusions was that conventional approaches in the transport sector appear to be narrow. Given also the anticipated launch of the Galileo satellite navigation system in Europe and the inclusion of 'smart' and 'green' policies as a major component of Horizon 2020, the European Union research programme 2014–20, this book is deemed to be a timely contribution to broaden and expand the debate about these pertinent issues.

One of the impressive aspects of this book is that it features chapters by experts across continents, providing a holistic view from a transport sector perspective which is beneficial for practitioners in the ICT sector. Each chapter has been written individually addressing the selected topic, yet all chapters highlight key challenges relevant to the overall book theme based on their respective context. Utilizing empirical evidence from a wide range of countries and transport modes, this volume summarizes major opportunities and threats which need to be considered locally and globally. As a result, the last chapters outline future objectives to be addressed by researchers, policy makers and practitioners in a range of sectors, based on the evaluation of reported experience and the assessment of transferability to other geographical areas to meet wider objectives.

This book is intended for anyone interested in understanding the opportunities, threats, principles and practical challenges faced by those who aim at better understanding the use and deployment of ICT in the transport sector. The book is of interest to academics, researchers, software developers, technology and innovation managers, multinational organisations, as well as local and national policy makers worldwide. The book's contents and bibliography provide up-to-date knowledge and expertise based on state-of-the-art research in Europe, America, Asia and Africa, but one need not be a professional or field expert to follow the main arguments and findings presented. This volume will be a valuable source of information for those interested in being at the forefront of the upcoming challenges in the evolving field of ICT for transport. This wander through the various threats and opportunities will assist them in making more informed decisions about the future use of ICT in transport and for the benefit of society.

Acknowledgements

Publication of this NECTAR series volume would not have been possible without the contribution, support and cooperation of several people around the globe. First of all, we would like to thank the late Professor Piet Rietveld, who urged us to embark on this endeavour despite the initial challenges. His broad overview of both transport and ICT issues, along with his calmness and experience laid the foundations for the book you are currently reading. He did contribute until the very end, so we are deeply indebted to him.

We are thankful to each one of the authors for agreeing to contribute to this book enthusiastically, sharing their expertise and findings to provide constructive recommendations for a diverse readership. This has been a long process where coordination and time planning have been constant challenges for both authors and editors. Along the same lines, we would like to express our gratitude to all the reviewers who wholeheartedly agreed to contribute their time and expertise to peer review chapters of this book. Dr Mark Andrejevic, Dr Costas Antoniou, Dr Susan Böhler, Dr Sigal Kaplan, Dr Stan Karanasios, Prof. Patricia Mokhtarian, Prof. Panos Papaioannou, Prof. Robert Paaswell, Prof. Miquel Angel Piera and Dr Dick de Ward acted as reviewers and we would like to thank each one individually for their contribution.

Additionally, we are grateful to Professor Aura Reggiani who also had an influential role in the early and final stages of this edited volume as the inaugurator of this NECTAR series and as the NECTAR Chair until 2013. Alan Sturmer and the Edward Elgar editorial team have also been very helpful with their flexible but at the same time professional and diligent approach which helped us deliver within the anticipated timeframe. Furthermore, we would like to thank NECTAR and the Regional Studies Association for the funds provided to host workshops and discuss the contents of selected chapters of this book.

1. Introduction: transport and ICT

Nikolas Thomopoulos, Moshe Givoni and Piet Rietveld

1 ICT EVOLUTION

Radical technological changes have occurred in the dawn of the twenty-first century, creating high expectations both for short-term and long-term evolutions in the use of information to improve trip accuracy and travel convenience within the transport sector. Information has become the global currency of this century and ICT (Information and Communication Technology) enables the distribution of information, facilitating both physical and digital accessibility. Coupled with increased mobility, the transport sector is constantly generating data through multiple devices, sensors and sources (Pelletier et al., 2011), an illustration of the unique feature of this era. Global mobile data traffic grew 81 per cent in 2013 and forecasts show that Africa and the Middle East are anticipated to grow – based on the compound annual growth rate – by 70 per cent whereas Central and Eastern Europe are anticipated to grow by 68 per cent (Cisco, 2014). So travellers and institutions have the ability to create value about themselves on their own, as members of organized communities or as part of the wider digital domain constantly producing or using data. It has been predicted that user-generated content and networking can act as democratizing forces in this context (Dutton, 2013).

Mobile phone penetration has reached peak levels in certain countries such as Italy, where almost every individual has two mobile phones, whilst smartphones are used by more than one out of three users in countries such as the UK, Sweden, Italy or Germany (OfCom, 2011). Smartphone penetration is over 70 per cent in both the United States and China (Phadke, 2013), with almost half of mobile phone users in developed countries including the United Kingdom, Germany, Italy and Australia using their phone to access the Internet. Over 1.2 million software applications (that is, apps) existed in June 2014 for each of the two major smartphone types, demonstrating an immense interest by users, businesses and governmental organizations for ICT through apps, social media and short

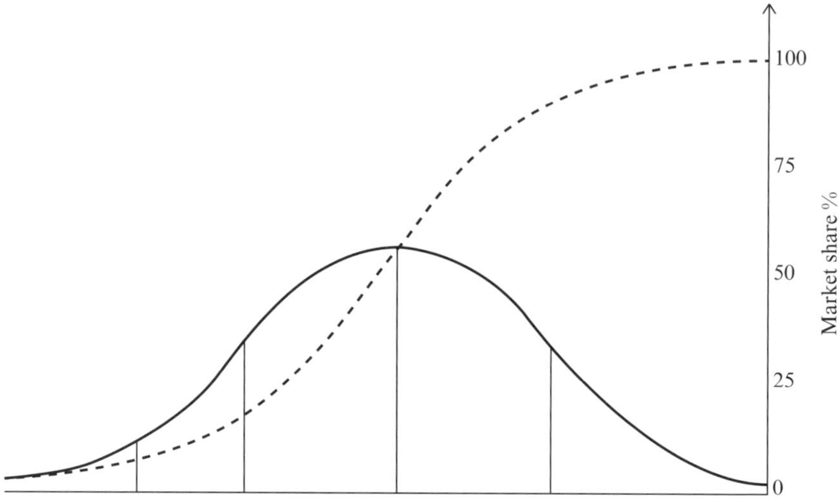

Notes: Solid line indicates market share of each of the 5 groupings. Broken line indicates cumulative adoption of an innovation as a proportion of the total market share. Graph courtesy of Marco Derksen.

Source: Hidalgo and Zeng (2013).

Figure 1.1 Five stages of diffusion of innovation

message services (SMS). On the other hand though, there are places in developing countries where five people share the same phone, highlighting the immense gap between societies and economies as well as the diverse rates of innovation adoption (Figure 1.1) in diverse contexts.

There are very few sectors other than transport where accurate and timely information is valued more. This is reflected also in the ever-growing number of transport-related smartphone apps as well as in the actual infrastructure installed in railway stations, airports, car parks or bus stops. More than 40 million travel information requests by smart phones are registered every month in Sydney (Berejiklian, 2014) and more than 70 per cent of Londoners use travel apps through their smartphone (Rode et al., 2014a) for instance.

This book is in line with the emerging interest in the intersection of ICT and transport, two related disciplines which have been brought closer recently due to increased infrastructure and technological developments. In conjunction with increased mobility worldwide and the widespread use of mobile technology and smartphones, this interest is expected to boom in the near future. The number of connected devices has been estimated to surpass 20 billion, with the potential to reach 50 billion by 2020, yielding net profits of more than US$1 trillion across a wide range of market sectors (Schindler et al., 2013).

However, technological advancements and the so-called Internet of Things are often accompanied by limitations, so identifying threats and mitigating risks is fundamental particularly at the early stages of development in order to fully exploit opportunities which may bring multiple benefits at various levels. Emerging hierarchies and practices have to be reviewed both locally and globally to counter rather than augment inequalities (Marwick, 2013). The budget constraints currently faced by certain developed countries pose more challenges since the design of efficient ICT for transport networks often comes at a high cost, but it is also an opportunity if it is able to replace, at least partly, investment in physical transport infrastructure, which is likely to be even more costly.

Consequently, it is crucial to draw attention and funds to the types of ICT for transport infrastructure needed the most at local, national or international levels in order to ensure that benefits reach a range of stakeholders. What types of ICT hardware, software or smartphone apps are able to engender interest from users while at the same time meeting environmental, social or economic targets? Are threats and opportunities common for all types of transport modes, for example motorway transport, public transport or non-motorized transport? Which are the key similarities and differences in the needs of developed and developing countries in the quest to establish effective ICT for transport applications? Since today's ICT will not be the ICT of tomorrow, which theoretical framework should be applied to ensure continuity? Such issues about ICT for transport focusing on people, processes, infrastructure and transport are addressed in this book, aligned with contemporary interest by global institutions such as the European Commission (Schindler et al., 2013) to ensure that relevant policies abide by established values and regulations whilst promoting a positive societal impact through technological innovation.

2 COMBINING ICT AND TRANSPORT: ICT FOR TRANSPORT

The link between ICT and transport started to evolve more than 30 years ago (Figure 1.2) and is anticipated to continue after 2020 with connectivity, digital demand and innovative solutions playing an instrumental role (Mohr et al., 2013). As presented in Chapter 11, the European Union

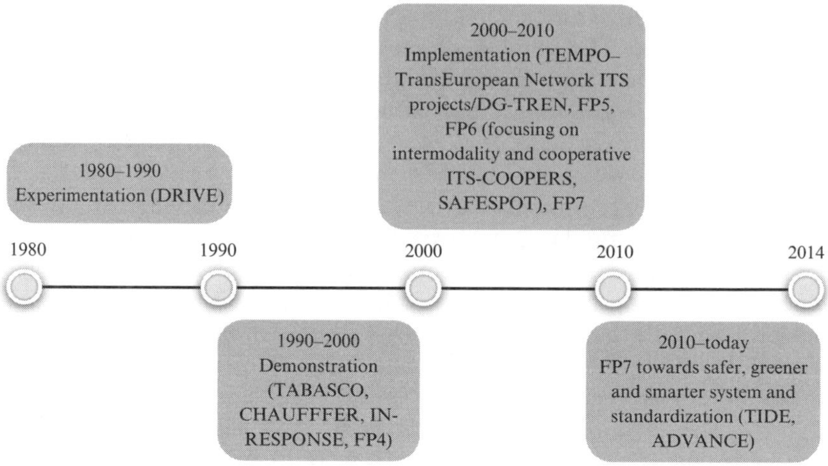

Source: Pitsiava-Latinopoulou (2014).

Figure 1.2 ITS timeline in Europe

established it further with Directive 2010/40/EU, which focuses on the deployment of Intelligent Transport Systems (ITS) for road transport and the interface with other transport modes (OJEU, 2010). Sustainable Urban Mobility Plans provide another platform to link ICT and transport in order to benefit users and meet urban policy objectives.

Global Interest

Global institutions have acknowledged during the past few years the importance of technology to meet economic, environmental and social objectives. Therefore the interest in ICT for transport and the role it is expected to play has become clear. This has been communicated via the terms *smart and green development*, used frequently after the Rio+20 summit (Ang and Marchal, 2013; Floater et al., 2014; OECD, 2010; UNEP, 2011; UN, 2012). These terms have been translated into World Bank policies varying from the freight and green trucks programme implemented in China (Mehndiratta et al., 2011) to the ICT in the context of green growth aiming at agriculture and climate change in Africa (Palmer, 2012). Similarly, the UN has stressed the potential of public transport to act as a smart and green solution, not only for developed but also for developing countries (UN, 2010) since it can reduce greenhouse gases (GHGs) emissions by 25 per cent and urban traffic fatalities by 15 per cent (Pourbaix, 2012). Technology transfer including ICT between developed and developing countries (EU, 2012) will enable the dispersion of benefits gained within the transport sector as well as in other sectors. In 2010 the European Union adopted a specific policy, *Europe 2020: a strategy for smart, sustainable and inclusive growth* (EU, 2010), which should meet relevant targets in the forthcoming years through EU headline targets, integrated guidelines, country-specific recommendations, devoted stability, convergence and national reform programmes (EC, 2010). In addition, 6.4 billion euros have been ring-fenced for smart, green and integrated transport through H2020, the respective EU programme for Research & Development (EC, 2014a) which aims to address uneven spatial dispersion of ICT innovation (EC, 2014b).

Smart Cities and the Use of Diverse Terminology

Nonetheless, certain barriers have not allowed a smooth integration of ICT and transport. To begin with, due to the use of diverse terms in the past decades to encompass the link between ICT and transport, a smooth integration of these sectors has not been permitted. Internet of Things, Web 3.0, Web 2.0, Cloud computing, ubiquitous computing, emerging technologies and Big Data[1] are not new terms to those within the computing and IT (Information Technology) sectors. Along with terms such as telematics, Intelligent Transport Systems, Information Technology and Transport or resilient systems, urban planners and transport practitioners have started to merge terminology, mainly within the intelligent, digital, electric, sustainable, future or smart cities context (Burdett and Rode, 2012; Komninos, 2002; 2008; 2014; Moir et al., 2014), which seems a fertile terrain where ICT for transport can be applied and flourish. Nonetheless, the terminology used is only the tip of the iceberg since competing priorities, diverse viewpoints and work patterns or funding allocation have hindered collaboration among researchers and practitioners within related disciplines and halting progress. The International Telecommunication Union (ITU) has recently looked into this evolving domain and attempted to shed light by reviewing terminology used in relation to smart cities,

---

1 Since no rigorous definition of Big Data exists, this term is defined based on Mayer-Schönberger and Cukier (2013): 'Big Data refers to things one can do at a large scale that cannot be done at a smaller one, to extract new insights or create new forms of value, in ways that change markets, organizations, the relationship between citizens and governments, and more'.

suggesting the following definition which highlights the key role of transport within this field:

> A smart sustainable city is a city that leverages the ICT infrastructure in an adaptable, reliable, scalable, accessible, secure, safe and resilient manner in order to: [. . .] Streamline physical infrastructure based services such as the transportation (mobility), water, utilities (energy), telecommunications, and manufacturing sectors (ITU, 2014)

Despite the different notions embraced by the contested concept of smart cities (Caragliu et al., 2011; Greenfield, 2013), it is widely accepted that certain cities have already managed to successfully integrate ICT and transport, enabling them to implement innovative solutions to address to some extent contemporary challenges such as congestion, traffic safety, emissions, cost of transport, and reliability.

Rio de Janeiro, for instance, has established an operations centre aiming at reducing emergency response time offering real-time updates to officials and citizens, whereas Singapore has agreed to offer know-how and support to build 100 smart cities in India, starting from the Delhi–Mumbai Industrial Corridor (DMIC). Here the merging of ICT and transport has a prominent role through computerized traffic systems, centralized traffic light management, the provision of real-time traffic information and public transport services at stations or on board. Along the same lines, the Connected Boulevard partnership between ICT corporations, urban solution providers and the city of Nice in France aims at proving the concept of the Internet of Everything within smart cities. Through the use of multiple sensors and networked devices, the aim is to generate data about everyday life such as smart traffic circulation, smart lighting, smart parking or environmental monitoring. Similarly, iPavement, which utilizes Wi-Fi and Bluetooth technology and has been tested on Via Inteligente in Madrid,[2] or solar roadways,[3] which have been developed as a prototype in the US and provide an interconnected network of data-generating sensors within cities or along motorways. This data will be owned and managed by the city or regional authorities in an attempt to generate innovative business models which would be able to finance such initiatives globally (Cisco, 2013).

Based on similar principles, the Oyster card introduced in 2003 offers

---

[2] More information is available at: http://www.ipavement.com/en.

[3] More information is available at: http://www.solarroadways.com, which also refers to the Department of Transportation SBIR grant (http://www.sbir.gov/sbirsearch/detail/355952).

improved reliability for travellers in London by allowing them to reach their destination through multimodal journeys using the underground, Docklands Light Rail, commuter trains, buses or river services while reducing transaction costs for operators and offering lower ticket costs for users. Such innovations have been feasible through the establishment of Transport for London (TfL), which benefits from the advantages of integrated transport authorities overseeing the provision of transport across administrative boundaries (Rode et al., 2014b). The next stage in London is anticipated to be the introduction of new ICT technologies via contactless payment for travel within Greater London as well as the integration of the public hire bicycle scheme for Oyster users.

Personalized travel schemes are also on the rise through technological advancements illustrated by GLIDE in Singapore, which offers more intersection crossing time for elderly or physically impaired pedestrians (Debnath et al., 2011) or by Smart Mobiliteit in Enschede, the Netherlands (Enschede Municipality, 2014), which supports self-motivated and rewarded travelling. Along with the first autonomous vehicle test in Europe to be realised in Ede-Wageningen, all these constitute best practice examples which can be assessed through existing measures such as the Readiness Guide (Berst et al., 2013). Such assessments contain the identification of a champion, the assembly of an appropriate team, the establishment of a larger vision for the wider area and the prioritization of objectives through predefined metrics.

Certain cities such as Amsterdam or Seoul have focused their smart evolution on sustainability, competitiveness and advanced ICT infrastructure (ITU, 2013) through specific ICT for transport initiatives (Figure 1.3). Yet even cities without a high quality ICT infrastructure can follow their own distinct path in yielding the benefits of ICT for transport by adopting initiatives gradually as part of a wider development plan (Figure 1.4). However, it is vital to circumvent locking in carbon-intensive and climate-vulnerable infrastructure in both developed and developing countries or to leapfrog by shifting investment towards sustainable transport that utilizes ICT according to its budget constraints and development stage (Ang and Marchal, 2013). The latter has been reflected in the policies adopted in the planned city of Masdar in Abu Dhabi, where technological optimism bias initially led the authorities to envisage a city with no private motor vehicles where transport would rely totally on Personal Rapid Transit. These plans have been revisited, though, to shift the focus more to electric and clean vehicles, mainly due to the high cost of introducing Personal Rapid Transit. The latter demonstrates an approach of adopting sustainable policies whilst avoiding carbon-intensive solutions in practice.

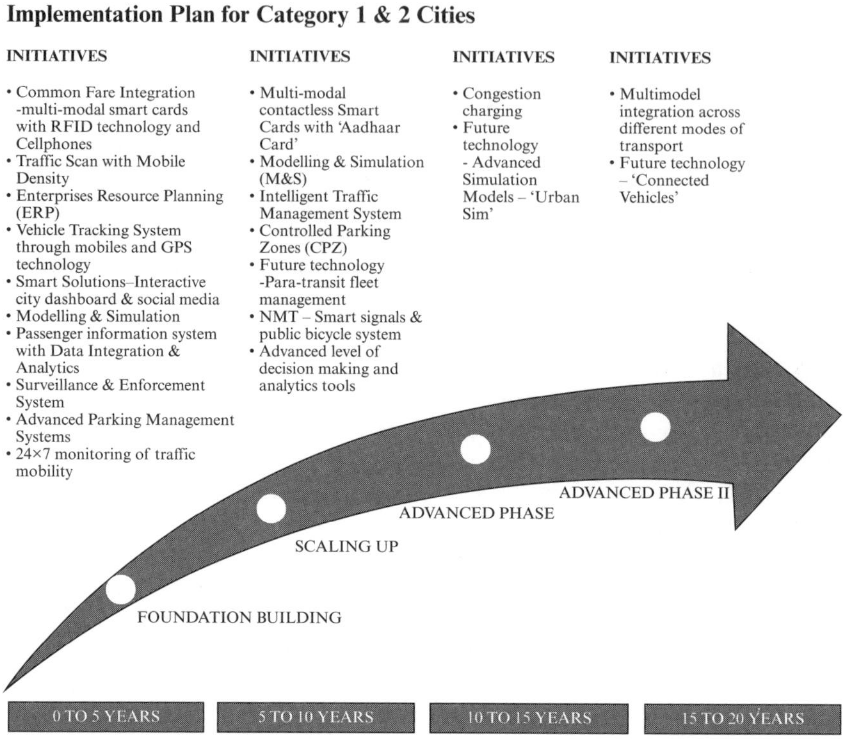

Source: GoI (2014, p. 36). Reproduced with permission.

Figure 1.3 Implementing ICT for transport in cities with high quality infrastructure available

ICT for Transport: a Contemporary Snapshot

Innovative transport schemes have been implemented globally since the nineteenth century when the London underground was inaugurated, but concepts such as shared mobility or multimodal integration have proliferated in the twenty-first century (Table 1.1), stressing the indisputable role of ICT for transport, also in the context of advancing sustainable transport. As demonstrated by Hidalgo and Zeng (2013), there is a strong correlation between the number of cities introducing after the year 2000 car sharing, bike sharing, smart cards and Bus Rapid Transit (BRT), and the technological infrastructure available in those cities. Expanding the use of digital information, for example through smart ticketing, has hugely benefited from the concept of shared mobility, which in turn has been

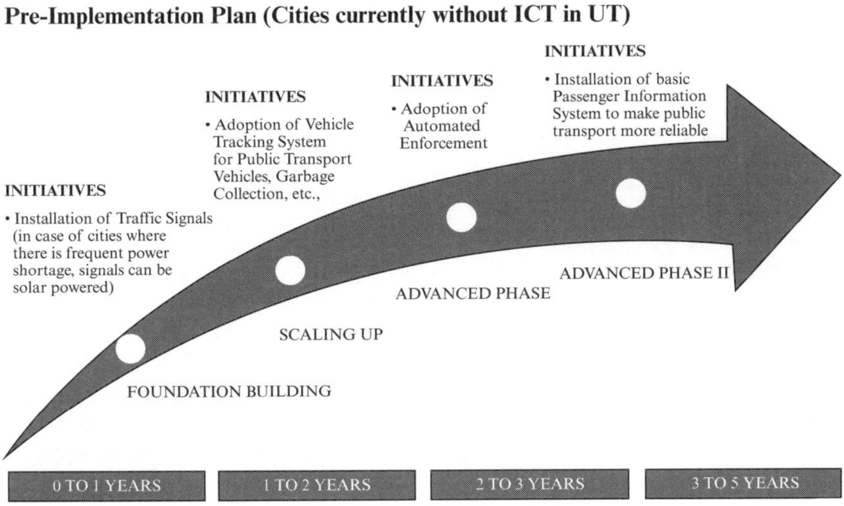

Source: GoI (2014, p. 36). Reproduced with permission.

Figure 1.4 Implementing ICT for transport in cities without ICT infrastructure

facilitated largely through the widespread use of integrated multimodal transport, ICT infrastructure, smartphones, social media and travel-related applications. As an illustration, the use of ICT has significantly altered public transport ticketing with contactless payment through smart cards. This is clearly reflected in Seoul, where 90 per cent of bus passengers and 75 per cent of underground passengers use smart cards to board buses and the underground respectively, offering travel benefits in the form of user convenience and reduced costs, as well as reduced transaction costs and time loss for operators (Pelletier et al., 2011). Notably, the more that cities implement such concepts, the more expertise and know-how will be exchanged between cities globally, reducing deployment costs whilst sharing best practice.

Equally, mobile Internet and smart phone applications have opened up major opportunities for shared mobility, which appears to be at the core of this revolution where new technologies are being developed to serve the needs of emerging urban economies. Those developments have been anticipated for more than half a century (Brotchie et al., 1991), but sharing a car, a bicycle or a taxi nowadays through apps such as Avego, BikeShare, Bixi, BlaBlaCar, car2go, Flywheel, Hailo, Lyft, Taxibeat, Uber, Zagster or Zipcar is not uncommon in several countries, from Brazil to India and from

Table 1.1 Sustainable transport concepts implemented worldwide

Area	Concept	Initial year of implementation (City)	Number of cities to date	Stage
Car restrictions & pricing approaches	Congestion pricing	1975 (Singapore)	6	Emerging
	Low emissions zones	2003 (Tokyo)	200+	Mainstream in Europe
	Vehicle quota systems	1990 (Singapore)	5	On the rise in China
Mass transit	Metro	1863 (London)	188	Mainstream in EU and North America, tipping in China
	Bus Rapid Transit (BRT)	1974 (Curitiba)	153	Tipping in Latin America and China; emerging in India
Shared mobility	Car sharing (2-way)	1987 (Lucerne, Zurich)	1000+	Mainstream in EU and North America; emerging in developing countries
	Bike sharing	1998 (Rennes)	≈500	Tipping in EU, the Americas and China
Urban design for access	Transit Oriented Development	Late 1800s and early 1900s (New York, London and others)	–	Mainstream in Europe, Japan, Hong Kong, Singapore; tipping in the US
	Car-free zones	1953 (Rotterdam)	360+ (213 in Europe)	Mainstream in European cities; tipping in North and Latin American cities
	Complete street	1971 (Portland)	[455] (in the US)	Tipping in the US

Table 1.1 (continued)

Area	Concept	Initial year of implementation (City)	Number of cities to date	Stage
Multimodal integration	Smart tickets	1989 (Zurich) 1992 (Oulu) 1996 (Seoul) 1997 (Hong Kong)	250+	Tipping in developed country cities and some emerging economies like China
	Information integration	2005 (Portland)	250+	Tipping in most developed country cities; on the rise in emerging economies

Source: Hidalgo and Zeng (2013).

Norway to Greece. Even large car manufacturers and car rental companies have realized the contemporary trends and have entered these evolving markets through Zipcar or car2go. Car-sharing schemes had grown to include 1.8 million registered users in 27 countries across all continents by the end of 2012 (Cohen, 2013) and new programmes have been launched in China, with electric vehicles also available for sharing in Hangzhou (Forbes, 2014). In this changing environment, Waze has revolutionized the way car drivers and commuters travel in countries such as Israel or the United States, and CityMapper has revolutionized the way public transport is used in Spain, the UK or the US. By contrast, bike-sharing programmes have proliferated across both developed and developing world cities, with 535 schemes operating in 49 countries in 2013, with Spain and Italy offering the most schemes in European cities while the Asia-Pacific region offers more than half of the global fleet (Larsen, 2013).

In this context, interesting views have started to emerge, such as the one expressed at the 12th annual Transport Practitioners' Meeting in London where the role and efficiency of TfL in developing and maintaining an online journey planner was questioned. Should public entities run such applications themselves in the future or should they merely collect and manage data offered to private entities who can then develop and maintain such ICT for transport applications more efficiently? The latter view raises several issues regarding data management, but it also brings up the potential role of cities and local authorities in acting as test beds for new ideas and technologies through living labs and public–private partnerships (PPPs). While existing applications like FixMyStreet, the first nationwide platform in the UK launched in 2007, which allows citizens to report problems such as potholes or broken streetlights directly in their local area, offer clear advantages in efficient management of resources and road space, others, like JustPark or ParkMe, offering private car parking space, raise questions about congestion, safety and tax revenue regulation.

Business Potential of Applications

Undoubtedly, there are high expectations for ICT advancements supporting business innovation to achieve improved efficiency in the transport sector, mainly to support management decisions (Kayikci and Zsifkovits, 2013) but also to entertain travellers through competitions or gamification of daily commuting aiming at community-wide environmental targets (Enschede Municipality, 2014; McCall et al., 2013). Nevertheless, mixed outcomes have been reported for logistics and taxi firms, partly also due to the limited amount of research devoted to the integration of diverse applications, the actual efficiency gains for hauliers, empirical findings on the adoption of new technologies by corporations, as well as the role of technology providers in the adoption and deployment of new technology (Perego et al., 2011; Sternberg et al., 2014). Although green logistics and green IT are areas of extensive focus for ICT developments within logistics, new areas are dominating the taxi sector (Frehe and Teuteberg, 2014; Murugesan and Gangadharan, 2012).

For instance, e-hailing apps on smart phones, which are used by travellers to call taxis, have gained significant popularity in China, reshaping the landscape of the taxi industry in large cities whilst attracting US$43 million in investment for this rapidly evolving sector in 2012–13. More than a third of the 68 000 taxis in Beijing have been using e-hailing apps since they became available in 2012. Analogous is the situation in other major cities such as Guangzhou, Hangzhou, Shanghai and Shenzen, where almost 90 per cent of the market based on daily apps orders has been led by two information technology firms, Tencent and Alibaba, which offer Didi Taxi and Kuaidi Taxi respectively (Jourdan, 2014). Interestingly though, Chinese authorities have stepped in, attempting to regulate this emerging market dominated by two leading players by imposing restrictions. Companies in Beijing are required to cooperate with the central telephone-based dispatch centre of the capital which forces them to pay them management fees. Likewise, Shanghai, which is a global financial centre, in March 2014 banned the use of e-hailing apps by taxi drivers during the rush hour and has entirely forbidden their use by private vehicles registered for car sharing (WSJ, 2014).

Such examples indicate a strong business opportunity regarding ICT for transport for software developers and private corporations, but at the same time they blow the whistle for a certain level of regulation to avoid unintended consequences and manage public open spaces. Opportunities have been cited for both larger and smaller cities to act as test beds for innovation through the implementation of living labs to reap financial, technological and social benefits (Hjalmarsson et al., 2012) particularly through platforms deployed for smart cities by ICT corporations such as the Operations Centre in Rio de Janeiro. Likewise, numerous opportunities affecting travel behaviour and the use of travel time are emerging through the introduction of technologies such as 4G-LTE[4] in trains, coaches and buses (Chang and Thomopoulos, 2014). The outstanding challenge is to test and establish viable business models that will benefit all stakeholders while offering wider societal benefits.

Big Data, Open Data and Negative Impacts

Consequently, it is evident that there is strong potential for positive impacts through ICT for transport applications by using resources more efficiently while meeting energy or environmental targets (Gavanas et al., 2014). The sensors currently used to support existing infrastructure and provide accurate data lead to the creation of large databases which will grow exponentially over time. Open Source code is a particular feature which can be turned into a prominent resource in the hands of local and national authorities. Combined with Big Data, there are good prospects for creating value through existing information and databases (Mayer-Schönberger and Cukier, 2013) which can then be converted into resources used to promote public and non-motorized transport along with sustainable development schemes or to aid in emergency situations, as a result of extreme weather events, for instance. The latter is a clear indication that ICT for transport can empower users, transcending existing geographical or social barriers as illustrated in Nairobi through the Digital Matatus project.[5] Nevertheless, as stressed in Chapter 6, one has to be cautious when unveiling the socio-spatial benefits of Big Data (Shelton et al., 2014).

---

[4] 4G-LTE refers to the 4th Generation of mobile telecommunication technology. It succeeds 2G and 3G and it precedes 5G. LTE (Long Term Evolution) refers to a technical standard of high-speed data transfer.

[5] More information is available at: http://www.digitalmatatus.com.

Since robust validation methods have yet to be developed, there is a risk of information asymmetry through the generation of large amounts of low-value information utilizing valuable resources through VGI (Voluntary Geographical Information) and crowdsourcing. Social media based on ICT may assist in reducing travel disruptions (Pender et al., 2014), but when employing Open Source software there is always a threat of data manipulation or misuse when internationally agreed data management procedures are not adhered to or are non-existent. Obviously, diverse approaches have to be adopted in developed and developing countries to minimize adverse consequences and the digital divide (Graham, 2013). Criticism of Big Data should not be ignored due to the ease and low cost of transport-related data collection using ICT. Despite it being a powerful tool, ICT does not constitute a magical elixir yet and entails risks due to common data biases, the universal need for theory supporting any analysis or the risk of spurious correlations (Economist, 2014). As a result, the use of Big Data or Open Data should not be discouraged, but at the same time decision makers should be made aware not only of the opportunities but also of the potential threats regarding issues such as surveillance, equity or wider social impacts of ICT for transport (Andrejevic, 2011; Castells et al., 2007; Greenfield, 2013; Kitchin, 2014; Pridmore, 2012; Thomopoulos and Karanasios, 2014; Wright and de Hert, 2012; Zurawski, 2011).

3 PREVIEW OF CHAPTERS

The first part of this book outlines the substantial changes which have occurred in the use of ICT for transport to influence travel behaviour, showcasing the societal impacts in urban areas. A review of theory and the use of incentives illustrates that altering travel behaviour is possible. However, to introduce a lasting impact and increase the use of public transport, a seamless ICT system is required. By reviewing the travel behaviour literature in Chapter 2, Ben-Elia and Avineri show how complex the behavioural response of travellers is to the provision of travel information. The provision of information is where most of the potential benefits from the use of ICT for transport probably lie, but securing such benefits may not be straightforward.

Subsequently, Pronello and Camusso explore in Chapter 3 the specific user requirements that may enhance public transport use and attract more urban travellers by altering their travel behaviour while meeting local or international transport and environmental policy targets. Their comparative approach allows a valuable insight into Italian and German public transport users, whereas their findings about travellers with reduced mobility are of particular interest for both policy makers and practitioners. Gyergyay showcases in Chapter 4 that it is possible to change travel behaviour through tangible rewards when promoting non-motorized transport. By focusing on school trips in an area on the outskirts of London, he demonstrates that ICT is not only able to facilitate the management and monitoring of incentives distribution, but also that it is able to foster walking, thus introducing a lasting effect even after such schemes end due to limited funding.

Part II constitutes the core of this book, demonstrating through surveys and empirical evidence that innovation is not an easy task and that ICT for transport is not a panacea. Yet a wide variety of perspectives are offered via diverse case studies aiming at pointing out best practice and lessons to be learnt. In Chapter 5 Zegras et al. present findings of a recent survey in a developing economy, namely an area of a global mega-city such as Mexico City. By researching personal security perceptions of public transport users they provide important insight in the ways in which ICT can enhance sustainable transport, particularly for vulnerable travellers. The findings are useful for those interested in public transport in developing economies as well as those interested in enhancing transport surveys through the use of ICT. On a different topic of high significance, Chapter 6 presents World Bank findings regarding geolocation data and crowdsourcing, both of which are intertwined with the emerging field of Voluntary Geographic Information (VGI). VGI offers the opportunity to interested individuals to contribute their knowledge and expertise in collaborative projects such as keeping maps up to date, and it benefits particularly from the use of ICT. Such efficient practices increase both bottom-up and top-down participation in decision making whilst contributing to the decline of the digital divide between societal groups and countries. Examples of such a divide become evident in Chapter 7, which touches upon the potential benefits of vehicle-to-vehicle (V2V) communication in a developed country setting. There is a threat that V2V or similar technologies will be accessible only to a few users, allowing them to use improved travel services whereas the majority of users will not have access to such technologies and the derived benefits. Through a simulation, Lu and Han underline the potential advantages of wider use of V2V communication for fuel efficiency and reduced travel time, but warn that these benefits in real life situations should be expected to be much lower than those obtained in 'clean', ideal conditions tested in simulations, especially those of uniform traffic flow. Implicitly, they raise the issue of equity by analysing the impacts of only a minority of road users being able to benefit from V2V systems. They also show that ICT can shorten travel time and fuel consumption, but usually not simultaneously. Along the same lines, Envall stresses the potential benefits of ICT for transport planning in Chapter 8 through the use of GPS data for cyclists in urban areas. Empirical evidence from Sweden and Slovenia indicates that cycle planning can be a high value activity for local transport authorities since ICT provides them with cost-efficient and accurate information about cycle routes and travel speeds.

It is the implementation of such innovative solutions that is needed to inform strategies, frameworks and policies globally. This is the theme of Part III, which ascertains the most relevant policy implications and makes relevant recommendations. Chapter 9 tries to expose some of the less explored aspects of the otherwise beneficial use of ICT for transport. Herzogenrath-Amelung et al. even refer to it as a 'mission creep' type of development, especially if ethical, social and legal considerations are not properly thought through. These are largely due to the marketing discourses within the development of ICT for transport applications that highlight benefits offered whilst downplaying or sometimes ignoring the risks involved. Privacy and surveillance issues raised by ICT applications in the transport domain are exemplified. Employing philosophical theories and applying them in London through the Oyster card case, they outline key challenges that need to be considered by decision makers when implementing 'smart' travel card applications. The prism under which ICT developments need to be discussed is not a single entity, namely a developer or a company, nor are these the responsibility of a single government agency. There is a network of actors, Communities of Interest (CoI), where these considerations must be accounted for. In Chapter 10 Angelidou et al. show the role of CoI in increasing the success of ICT development, and call for policy makers to provide the support for such CoI to function better. Along the same line of argument, Tyrinopoulos et al. present in Chapter 11 a long list of successful implementations of ICT for transport applications across the EU (European Union). Evidence is critically analysed and categorized and this chapter also maps the reference to ICT over time in EU transport policy reports.

As an epilogue, the concluding chapter summarizes findings and views reported within this book to build on potential synergies, minimizing contradictions. Thus, it provides a framework approach which may be used by policy makers, researchers or practitioners worldwide to guide decision making at a local or global level.

BIBLIOGRAPHY

Andrejevic, M. (2011) Surveillance and alienation in the online economy, *Surveillance & Society*, 8(3), pp. 278–87.

Ang, G. and Marchal, V. (2013) Mobilising private investment in sustainable transport: the case of land-based passenger transport infrastructure, *OECD Environment Working Papers No. 56*.
Banister, D. (2008) The sustainable mobility paradigm, *Transport Policy*, 15(2), pp. 73–80.
Banister, D. and Hickman, R. (2006) How to design a more sustainable and fairer built environment: transport and communications, IEEE Proceedings of the Intelligent Transport System, 153(4).
Berejiklian, G. (2014) Number of customers using real time transport apps surges. Minister for Transport, Chippendale, NSW, 5 March.
Berst, J., Enbysk, L. and Williams, C. (2013) Smart cities readiness guide, Smart Cities Council, p. 27. Retrieved 22 January 2014 from: http://www.rinnovabili.it/wp-content/uploads/2013/12/SmartCitiesCouncil-ReadinessGuide.pdf.
Brotchie, J., Batty, M., Hall, P. and Newton, P. (1991) Cities of the 21st century: new technologies and spatial systems, Melbourne: Longman Cheshire.
Burdett, R. and Rode, P. (eds) (2012) The electric city, in R. Burdett and P. Rode, Urban Age Electric City conference, London 6–7 December 2012. The London School of Economics and Political Science, Alfred Herrhausen Society, London.
Caragliu, A., Del Bo, C. and Nijkamp, P. (2011) Smart cities in Europe, *Journal of Urban Technology*, 18(2), pp. 65–82.
Castells, M., Fernández-Ardèvol, M., Linchuan Qiu, J. and Sey, A. (2007) Mobile communication and society: a global perspective, Cambridge, MA: MIT Press.
Chang, T.-P. and Thomopoulos, N. (2014) The potential impact of 4G on a main rail route in the UK, paper presented at the PTRC annual Transport Practitioners Meeting, 2–3 July, London.
Cisco (2013) Cisco and Nice unveil Connected Boulevard, press release. Retrieved 17 March 2014 from: http://files.shareholder.com/downloads/CSCO/3441683826x0x671232/b6631003-51b8-415a-8b25-3f51695a3f43/CSCO_News_2013_6_17_Press_Releases.pdf.
Cisco (2014) Cisco Visual Networking Index: global mobile data traffic forecast update 2013–2018. Retrieved 21 May 2014 from: http://www.cisco.com/c/en/us/solutions/collateral/service-provider/visual-networking-index-vni/white_paper_c11-520862.html.
Cohen, A. (2013) Innovative Mobility Carsharing Outlook: carsharing market overview, analysis, and trends – Summer 2013. Retrieved 21 May 2014 from: http://www.innovativemobility.org/publications/Carsharing_Innovative_Mobility_Industry_Outlook.shtml.
Debnath, A., Haque, M., Chin, H. and Yuen, B. (2011) Sustainable urban transport: smart technology initiatives in Singapore, *Transportation Research Record*, 2243, pp. 38–45.
Dutton, W. (2013) Running in social media circles. Review of Alice E. Marwick's 'Status update: celebrity publicity and branding in the social media age', New Haven, CT: Yale University Press. *Science*, 342(933).
EC (2010) Europe 2020 timeline. European Commission, 25 January. Retrieved 21 January 2014 from: http://ec.europa.eu/archives/growthandjobs_2009/pdf/annex3_en.pdf.
EC (2014a) Smart, green and integrated transport, European Commission, Horizon 2020. Retrieved 21 May 2014 from: http://ec.europa.eu/programmes/horizon2020/en/h2020-section/smart-green-and-integrated-transport.
EC (2014b) Sixth report on economic, social and territorial cohesion, July,

DG Regional and Urban Policy, Brussels. Retrieved 1 August 2014 from: http://ec.europa.eu/regional_policy/sources/docoffic/official/reports/cohesion6/6cr_en.pdf.
Economist (2014) The backlash against Big Data, Economist explains, 20 January. Retrieved 21 May 2014 from: http://www.economist.com/blogs/economist-explains/2014/04/economist-explains-10.
Enschede Municipality (2014) SMART application. Retrieved 3 June 2014 from: http://www.smartintwente.nl.
EU (2010) Europe 2020: a European strategy for smart, sustainable, and inclusive growth, European Commission COM(2010) 2020. Retrieved 21 January 2014 from: http://ec.europa.eu/eu2020/pdf/COMPLET%20EN%20BARROSO%20%20%20007%20-%20Europe%202020%20-%20EN%20version.pdf.
EU (2012) Conclusions on Rio+20: Outcome and follow-up to the UNCSD 2012 Summit, 3194th ENVIRONMENT Council meeting, Luxembourg, 25 October. Council of the European Union. Retrieved 21 January 2014 from: http://www.consilium.europa.eu/uedocs/cms_data/docs/pressdata/en/envir/133219.pdf.
Floater, G., Rode, P., Zenghelis, D., Ulterino, M., Smith, D., Baker, K. and Heeckt, K. (2014) Copenhagen – green economy leader report, London: LSE Cities.
Foray, D., Goddard, J., Beldarrain, X., Landabaso, M., McCann, P., Morgan, P., Nauwelaers, C. and Ortega-Argiles, R. (2012) Guide to research and innovation strategies for smart specialisations (RIS3), Luxembourg: Publications Office of the European Union.
Forbes (2014) Kandi crush: an electric-car vending machine from China could upend the auto industry. Retrieved 3 June 2014 from: http://www.forbes.com/sites/markrogowsky/2013/12/28/kandi-crush-an-electric-car-vending-machine-from-china-could-upend-the-auto-industry/.
Frehe, V. and Teuteberg, F. (2014) The role of ICT in green logistics: a systematic literature review, in B. Funk et al. (eds), Information technology in environmental engineering, Berlin and Heidelberg: Springer, pp. 53–65.
Gavanas, N., Kouridis, C., Tsakalidis, A., Verani, E., Papageorgiou, T. and Blatsis, E. (2014) ICT platform for road users' real-time information and guidance based on environmental criteria, paper presented at the Regional Studies Association (RSA) Research Network workshop on Theory and practice of using ICT to facilitate smart and green regional development, 5–6 June 2014, Thessaloniki, Greece.
Givoni, M. and Banister, D. (eds) (2013) Moving towards low carbon mobility, Cheltenham, UK and Northampton, MA, USA: Edward Elgar Publishing.
GoI (2014) India transport report: moving India to 2032, Volume I – Executive summary, Government of India, New Delhi, Routledge. Also available at: planningcommission.nic.in/reports/genrep/NTDPC_Vol01.pdf.
Graham, M. (2013) Geographies of information in Africa: Wikipedia and uneven user-generated content, R-Link, January–March 2013, pp. 40–41.
Greenfield, A. (2013) Against the smart city, New York: Do Projects.
Hidalgo, D. and Zeng, H. (2013) On the move: pushing sustainable transport from concept to tipping point. Embarq blog. Retrieved 25 April, 2014, from: http://thecityfix.com/blog/on-the-move-pushing-sustainable-transport-concept-tipping-point-dario-hidalgo-heshuang-zeng/.
Hjalmarsson, A., Thomopoulos, N., Hodgson, F., Klok, E., Bjilsma, M. and Dethmers, V. (2012) D5.3 Business aspects, Sustainable Social Networking

Services for Transport (SUNSET) Deliverable, FP7 no.270228, DG Information Society & Media, Brussels.
ITU (2013) Smart cities: Seoul – a case study, Geneva: International Telecommunication Union. Retrieved 25 April 2014 from: http://www.itu.int/dms_pub/itu-t/oth/23/01/T23010000190001PDFE.pdf.
ITU (2014) WG1 Smart Sustainable Cities – Analysis of Definitions, Technical report SSC-0100-rev-3, Telecommunication Standardization Sector, Genoa, 19–20 June.
Jourdan, A. (2014) Shanghai cracks down on booming taxi app market, *Reuters*, 26 February. Retrieved 21 May 2014 from: http://www.reuters.com/article/2014/02/27/us-china-taxiapps-idUSBREA1Q03V20140227.
Kayikci, Y. and Zsifkovits, H. (2013) Successful ICT integration in transport collaboration, in T. Blecker et al. (eds), Pioneering solutions in supply chain performance management, vol. 17 – Supply chain, logistics and operations management, Cologne: Josef EUL Verlag.
Kitchin, R. (2014) The data revolution: big data, open data, data infrastructures and their consequences, Thousand Oaks, CA: Sage.
Komninos, N. (2002) Intelligent cities: innovation, knowledge systems and digital spaces, Abingdon: Routledge.
Komninos, N. (2008) Cities and globalisation of innovation networks, Abingdon: Routledge.
Komninos, N. (2014) The age of intelligent cities: smart environments and innovation-for-all strategies, Abingdon: Routledge.
Larsen, J. (2013) Bike-sharing programs hit the streets in over 500 cities worldwide, Earth Policy Institute. Retrieved 3 June 2014 from: http://www.earth-policy.org/plan_b_updates/2013/update112.
Marwick, A. (2013) Running in social media circles, *Science*, 342(6161), pp. 933–4.
Mayer-Schönberger, V. and Cukier, K. (2013) Big Data: a revolution that will transform how we live, work and think, New York: Houghton Mifflin Harcourt.
McCall, R., Koenig, V. and Kracheel, M. (2013) Using gamification and metaphor to design a mobility platform for commuters, *International Journal of Mobile Human Computer Interaction*, 5(1), pp. 1–15.
Mehndiratta, S., Fang, K. and Darido, G. (2011) World Bank case studies on green trucks/freight transport initiatives: China case study, May. Retrieved 25 April 2014 from: http://sustainabledevelopment.un.org/content/dsd/csd/csd_pdfs/csd-19/learningcentre/presentations/May%209%20pm/2%20-%20Georges%20Darido%20-%20World%20Bank%20Green%20Freight%20Cases.pdf.
Mohr, D., Mueller, N., Krieg, A., Gao, P., Kaas, H-W., Krieger, A. and Hensley, R. (2013) The road to 2020 and beyond: what's driving the global automotive industry, Advanced industries. Stuttgart: McKinsey & Company.
Moir, E., Moonen, T. and Clark, G. (2014) What are future cities? Origins, meanings and uses. Foresight Future of Cities project, Future Cities Catapult, Government Office for Science.
Murugesan, S. and Gangadharan, G. (2012) Harnessing Green IT: principles and practices, IEEE, New York: Wiley.
O'Brien, C. (2013) Towards a dynamic and trustworthy Internet of Things. RAND Europe Research Brief. Retrieved 21 May 2014 from: http://www.rand.org/content/dam/rand/pubs/research_briefs/RB9700/RB9742/RAND_RB9742.pdf.
OECD (2010) Greener and smarter: ICTs, the environment and climate change, Paris: OECD Publishing.

OfCom (2011) The communications market report, London: Office of Communications, HMC.

OJEU (2010) Directive 2010/40/EU of the European Parliament and of the Council of 7th July 2010 on the framework for the deployment of Intelligent Transport Systems in the field of road transport and for interfaces with other modes of transport. Official Journal of the European Union L207/1, 6 August. Retrieved 25 April 2014 from: http://eur-lex.europa.eu/LexUriServ/LexUriServ.do?uri=OJ:L:2010:207:0001:0013:EN:PDF.

Palmer, N. (2012) ICT and agriculture in the context of green growth, e-Sourcebook, Forum, 2 March.

Pelletier, M-P., Trépanier, M. and Morency, C. (2011) Smart card data use in public transit: a literature review, Transportation Research Part C: Emerging technologies, 19(4), pp. 557–68.

Pender, B., Currie, G., Delbosc, A. and Shiwakoti, N. (2014) Social media use during unplanned transit network disruptions: a review of literature, *Transport Reviews*, 34(4), pp. 501–21.

Perego, A., Peroti, S. and Mangiracina, R. (2011) ICT for logistics and freight transportation: a literature review and research agenda, *International Journal of Physical Distribution & Logistics Management*, 41(5), pp. 457–83.

Phadke, S. (2013) The Asian mobile consumer decoded. Retrieved 21 May 2014 from: http://www.nielsen.com/us/en/newswire/2013/the-asian-mobile-consumer-decoded0.html.

Pitsiava-Latinopoulou, M. (2014) ITS towards sustainable mobility in urban areas. Keynote speech at Regional Studies Association (RSA) Research Network workshop on Theory and practice of using ICT to facilitate smart and green regional development, 5–6 June 2014, Thessaloniki, Greece. Available at: http://www.regionalstudies.org/networks/network/using-ict-to-facilitate-smart-and-green-regional-growth.

Pourbaix, J. (2012) Towards a smart future for cities, *Journeys*, May, pp. 7–13.

Pridmore, J. (2012) Consumer surveillance: context, perspectives and concerns in the personal information economy. In K. Ball et al. (eds) *Routledge Handbook of Surveillance Studies*, Abingdon: Routledge.

Rietveld, P. and Vickerman, R. (2004) Transport in regional science: The 'death of distance' is premature, *Papers in Regional Science*, 83(1), pp. 229–48.

Rode, P., Hoffmann, C., Kandt, J., Graff, A. and Smith, D. (2014a) New urban mobility in London and Berlin. Forthcoming. LSE Cities and InnoZ.

Rode, P., Floater, G., Thomopoulos, N., Schwinger, P., Mahendra, A. and Fang, W. (2014b) Accessibility in cities: transport and urban form. New Climate Economy Cities Paper 03. LSE Cities. London School of Economics and Political Science.

Schindler, H.R. et al. (2013) Towards a dynamic and trustworthy Internet of Things. Research Report, Rand, available at http://www.rand.org/pubs/research_briefs/RB9742.html.

Shelton, T., Poorthuis, A., Graham, M. and Zook, M. (2014) Mapping the data shadows of hurricane Sandy: uncovering the sociospatial dimensions of 'Big Data'. *Geoforum*, 52, pp. 167–79.

Sternberg, H., Prockl, G. and Holmstrom, J. (2014) The efficiency potential of ICT in haulier operations, *Computers in Industry*, 65(8), pp. 1161–8.

Thomopoulos, N. and Karanasios, S. (2014) Social impacts of Information and Communication Technologies in transport, in M. Garrett (ed.), Encyclopedia

of Transportation: Social Science and Policy, Thousand Oaks: Sage, pp. 575–8.
UN (2010) Public transport – the smart, green solution, Financing public transport, United Nations Centre for Regional Development, 5th Regional Environmentally Sustainable Transport (EST) Forum in Asia, 23–25 August, Bangkok, Thailand.
UN (2012) The future we want. United Nations Conference on Sustainable Development, Rio+20. Retrieved 21 May 2014 from: http://www.uncsd2012.org/content/documents/727The%20Future%20We%20Want%2019%20June%201230pm.pdf.
UNEP (2011) Towards a green economy: pathways to sustainable development and poverty eradication, United Nations Environment Programme.
Wright, D. and de Hert, P. (eds) (2012) Privacy Impact Assessment, the Netherlands: Springer.
WSJ (2014) Shanghai bans taxi apps during rush hour, *Wall Street Journal*, 27 February. Retrieved 21 May 2014 from: http://blogs.wsj.com/chinarealtime/2014/02/27/shanghai-bans-taxi-apps-during-rush-hour/.
Zurawski, N. (2011) Local practice and global data: loyalty cards, social practices, and consumer surveillance, *The Sociological Quarterly*, 52, pp. 509–27.

PART I

Changing travel behaviour through ICT for transport

2. The opportunities and threats of travel information: a behavioural perspective

Eran Ben-Elia and Erel Avineri

1 INTRODUCTION

In the developed world, we are in the midst of a technological takeover. Pervasive Information and Communication Technologies (ICT) are gradually infiltrating almost every computerized system used for human purposes. A recent trend is the incorporation of ICT in urban systems within the scope of Smart Cities, which include transport systems (Schaffers et al., 2011). ICT has been associated with a myriad of complex impacts on human activity and travel patterns. More than twenty years ago, Salomon (1986) and Mokhtarian (1990) described four mechanisms by which telecommunications (the term used for describing ICT at the time) were hypothesized to influence travel: substitution (or replacement of physical travel with virtual activity such as e-working, e-shopping, etc.); generation (or complementarity, or new trip stimulation); modification (of a trip that would have occurred anyway); and neutrality (or no influence). Twenty years on, this four-part definition has undergone some amendments, including multitasking, reallocation of resources (money and time) and activity (rather than trip) modification in association with ICT use (Mokhtarian, 2009).

Where does travel information fall within this complex framework of travel and ICT? Travel information is created, in part, from aggregated travel behaviour patterns emerging from signals and other data acquired from various ICT sources (such as floating data devices and mobile phones); at the same time travel information influences travel behaviours, with different technologies having different impacts on behaviours. Travel information has also evolved in parallel with the development of ICT and the media and devices it works through. Before the age of telecommunications, travel information was a limited affair based on a combination of formal information supplied by the transportation operators (usually in a physical form such as notice boards and road signs) and word of mouth (seeking advice from peers, relatives, friends or neighbours on how best to travel). Even today, in the presence of real-time and online travel information, word of mouth is still widespread, though it often takes more technologically advanced forms such as social media outlets (Bartle et al., 2013). Uncertainty about travel times was a common affair and being in a certain time and place was very much the determining factor as to whether travel information was accessible or not. Possibly the most common source of information was experiential, acquired through learning over time by repeated travelling experience. All this has changed with the start of the information age. Through their wired PCs, travellers could access and acquire information available online on web-based portals on the Internet or while on the road via electronic variable message signs (VMS). The first generation of travel planners (for example OV9292 in the Netherlands or Transport Direct in the UK) provided a source for pre-trip information, allowing travellers to make plans before departure. Physical presence at the source of information was critical to its availability. The phasing in of mobile and wireless communication technologies removed this barrier, allowing connected travellers to access real-time information anytime and anywhere while stationary or on the move through modified mobile platforms using personal specialized nomadic devices such as laptops, mobile phones and sat-navs. The rise of the app and the proliferation of high-speed Internet and smartphones have brought a bonanza of travel information at the touch of a button. Today, the generation and provision of travel information are no longer restricted to public agencies and road authorities and their designated infrastructures (for example, embedded loops under road surfaces or GPS tracked public transport vehicles): private enterprise is now leading the way forward with a multitude of apps providing information simply by tapping in to the source of the crowd (hence the term 'crowdsourcing') to extract information and communicate it back to their customers. The lucrative aspects of this new market have been elucidated with the purchase by Google of the social navigation start-up company Waze, which provides personalized real-time routing information, for a colossal sum of more than US$1.3 billion.[1] Nowadays we are probably for the first time over-informed, bombarded by many sources of information, both formal and informal. The main issue facing travellers today is integration of all these sources to make one user-friendly accessible pot where information can be easily found.

Despite this tremendous leap in technological capabilities on the supply

[1] 'Google Maps and Waze, outsmarting traffic together', 11 June, accessed 3 March 2014 at: http://googleblog.blogspot.co.il/2013/06/google-maps-and-waze-outsmarting.html.

side of travel information, very little is still known about ICT influences on the demand side – that is, how does information influence the behaviour of individual travellers, and what then are the systemic consequences for the performance of our transport networks? In fact, information provision is seen by many practitioners as a kind of panacea for solving many of our transport-related problems, namely traffic congestion. It is commonly believed that if only travellers received the right kind and amount of information then they would spontaneously react accordingly and the transport system would reorganize efficiently as a result (Levinson, 2003). This rather naive view negates much of the positive-empirical behavioural sciences and behavioural economics literature of the last three decades. This less known and less explored side of travel information is the focus of this chapter. We will start by developing the behavioural foundations of human choice under conditions of uncertainty which are associated with the travelling experience. Next, the empirical literature relating to behaviour and information is reviewed. We will finish with conclusions for further research and prospects for the future.

2 BEHAVIOURAL FOUNDATIONS

2.1 Rational Choice and Decisions under Uncertainty

Traditionally, transport theory has been developed with strong behavioural links to neoclassical economics. Specifically, the concept of *homo-economicus*, or economic man, has been a strong theme in travel demand models from their early origins in the 1960s. The fundamental theory behind economic man is rational choice. Rational choice is a well-established decision-making paradigm. In its most basic description an individual is said to behave rationally in a decision context if, when given information on all the attributes, costs and benefits of all the possible relevant alternatives she could choose from, she will choose the alternative which maximizes her cognitive hedonic target function – denoted as *utility*. Utility is an abstract concept, defined by the nineteenth-century philosopher Jeremy Bentham to measure hedonic pleasure, and later adopted by economists in consumer theory from where it has rolled towards transport science. In economics, utility describes a mathematical relationship of ordinal scale between consumption of goods and pleasure.[2]

[2] Rather than a measure of hedonic pleasure, the term 'utility' is commonly used nowadays as a measurement of 'helpfulness' or 'net benefit' (and in some contexts of individual decision making, 'utilitarian' and 'hedonic' are even seen as contrasted terms).

Utility's carriers are end-states, that is, more (of a good thing) is better than less. Moreover, utility is assumed to have the property of diminishing sensitivity, that is, more is better than less but much more is not that much better than more, or in mathematical terms, the utility function is concave. Utility theory has other specific axioms, like transitivity in alternative ranking. In transport science, utility is often related to the attributes of travel alternatives, for example, the cost and time of travelling by car or by train, but also the level of comfort and its environmental friendliness can be considered as attributes and has been applied within the scope of Random Utility Models (RUM), using discrete-choice specifications to estimate travel behaviour models, for example – mode and route choice (Ben-Akiva and Lerman, 1985; Train, 2002). These models have the advantage of being parsimonious and their flexible error structures allow the accommodation of a large range of unknown factors that influence individual choice (Prashker and Bekhor, 2004).

Uncertainty is a condition associated with choice environment where the outcomes of a choice are not known and probabilities are assigned to possible outcomes. It implies that the choice maker does not possess perfect information. Uncertainty is very much a basic characteristic of everyday travel where the traveller does not know for certain, once departing, the exact time of arrival (or other attributes of her journey). When assumed to behave rationally, under uncertainty a choice maker will multiply the utility of each alternative by its probability to arrive at an expected utility value of which she will prefer the highest. This natural extension of rational choice is known as Expected Utility Theory (EUT), first introduced by Bernoulli in 1738, and later extended formally by Von Neumann and Morgenstern (1944) and Luce and Raiffa (1957). A resulting behaviour from EUT axioms is that a rational choice maker will always be risk averse, that is she will prefer a more certain outcome compared to a gamble of equivalent expected value (for example, £500 is better than £1000 if a tossed coin falls on heads and 0 otherwise).

2.2 Bounded Rationality and Heuristics

Although characterized by mathematical elegance, a main critique of rational choice, in both general and transport contexts, is its normative nature (Gärling and Young, 2001). Rational choice ascribes how individuals ought to behave; it does not, however, describe how individuals actually behave, that is, its positive value is more limited. Empirical research conducted in behavioural sciences over more than half a century has found systematic violations of the fundamental axioms of rational choice and EUT in lab-based experiments. Some of this criticism focuses on the limitations of human cognition. Simon (1982) states that individuals are rationally bounded, that is, they cannot acquire and process complete information on all alternatives and all their attributes, and are incapable of making complicated calculations. Instead, they have limited choice sets and they tend to focus on the most important attributes, apply simple heuristics to solve complicated problems and will compromise on the alternative which is both satisfactory and sufficient – denoted as satisficing. Unlike the script-based optimization algorithm of rational choice, human cognition is composed of an adaptive toolbox of heuristics. Heuristics share key traits: they are fast, frugal and computationally cheap, rather than consistent, coherent and general. Heuristics arise from an evolutionary necessity, allowing the organism to respond rapidly to risks to survival in the environment. Different heuristics are applied for solving different problems in different choice environments, that is, they are context dependent (Gigerenzer and Selten, 2001). Unlike rational choice, which is normative and universal, the realm of heuristics is empirical and messy – different behavioural generalizations often give rise to contradictory behaviour when applying different heuristics. Kahneman (2011) coined the term 'thinking fast and slow' to illustrate the two cognitive systems the human mind possesses: one is slow, thorough, allowing complex problems to be solved in a reasoned manner, but has a high energy tag in consumed calories, and so, prone to laziness, that is, it tries to avoid being employed as much as possible. The second is blazing fast, energy saving, working automatically, often unconsciously and intuitively, and so, prone to making mistakes. This automatic system is the applier of most heuristics. Here we will discuss several well documented generalizations and effects that are of importance for understanding how travel information shapes travel behaviour.

Hot stoves, reinforced learning and experiential information

Based on the observation that most people refrain from touching for a second time a hot stove's door, Denrell and March (2001) and Denrell (2007) demonstrate that this behaviour is a natural consequence of the inherent asymmetry between the effect of good and bad experiences. In line with the 'classic' *law of effect* (Thorndike, 1898), good outcomes are expected to increase the probability that a choice will be repeated and facilitate the collection of information concerning the alternative that has yielded the good outcome. Bad outcomes are expected to reduce the probability that the choice will be repeated, and impair the collection of information concerning the alternative that has yielded the bad outcome. As a result, the impression of bad outcomes lasts longer than the effect of good outcomes. Studies based on reinforced learning also assert that increasing the noise level in the choice environment makes people more confused, impeding learning, and results in choices moving closer to random choice (i.e. indifference). Known as the 'payoff variability effect', it occurs when the choice maker relies on feedback from past experience (Erev and Barron, 2005) to form knowledge on the distribution of outcomes. Experience helps us learn from our mistakes, but, at the same time, how we learn depends to a large degree on how much information we acquire. As information acquisition is a function of sampling experiential information from our memory, quite biased estimates are typical, as described by Tversky and Kahneman (1974).

Framing, loss aversion and descriptional information

Kahneman and Tversky (1984) assert that choice makers frame uncertain outcomes as gains or losses relative to a reference point and not according to end-states as the classic interpretation of EUT suggests. Prospect Theory, initially developed by Kahneman and Tversky (1979) and later extended by Tversky and Kahneman (1992), tries to positively explain people's behaviours in situations involving one-shot decisions, based on information describing available alternatives. Like EUT, Prospect Theory assumes diminishing sensitivity. However, contrary to EUT, which assumes constant risk aversion, the theory predicts a choice maker will usually reveal risk-averse behaviour in the domain of gains, whereas risk-seeking behaviour will be observed in the domain of losses.[3] This effect, known as 'loss aversion' (Kahneman et al., 1991), happens since a loss looms larger than any equivalent gain. Loss aversion is a main motivation for people tending to stick to their current known state (status-quo bias), since the disadvantages of moving to an ambiguous situation appear greater than the advantages. Interestingly, experience-based studies refute some of the findings of description-based studies. In particular, the basic assumptions of Prospect Theory do not necessarily hold when moving from one-shot to repeated sequences of choices. Barron and Erev (2003) found risk attitude reversals when feedback is introduced in repeated choice experiments. Contrary to one-shot decisions, they assert that in repeated choice situations with feedbacks, participants tend to avoid risks when faced with losses and accept more risks when faced with prospective gains. Erev et al. (2008) further assert that in repeated choice situations with immediate feedback, behavioural tendencies previously

---

[3] To illustrate this peculiarity, consider the choice problem 'choose between: (A) there is an 80% chance of losing $4000, and (B) lose $3200 with certainty (100%)'. Although the expected values are identical, most people would prefer the gamble or risky prospect because it provides a chance of avoiding the unpleasantness of a sure loss.

related to loss aversion in decisions from experience are better described as consequences of diminishing sensitivity to absolute payoffs.

Overweighting and underweighting of rare events

EUT assumes people treat probabilities linearly. However, experiments show that descriptional information will also be associated with overweighting of small probabilities (Kahneman and Tversky, 1979). This means people will view possible outcomes as more likely than they really are. Consider the appeal of lottery tickets. The information provided describes a huge prize that the ticket buyer can potentially win. Yet the probability of such an event happening is miniscule, and still, people buy lottery tickets even though the expected value of winning is less than the price of the ticket. Conversely, with experiential information, underweighting is common (Camilleri and Newell, 2011; Hertwig et al., 2004). Think about insurance policies that once required you to remove the front panel of your CD player from your car when not present in the vehicle. Most people will have done so for a few days but soon they realized that the chance their car will be burgled is quite low. In fact, their risk perception became too low, leading them to ignore an objective risk. Sure enough, once in a while a burglary does occur, but this is already too late for the policy holder who is left with a void insurance due to her personal negligence.

The lure of affect

Affect relates to things we enjoy and like. However, what we like can also cloud our better judgement and often leads us to poor choices. Slovic et al. (2007) suggest the affect heuristic lures people into risky behaviours by the prospects of fun and excitement. Think about young smokers who should be well aware of the risks they are taking – both financially and health-wise – and yet, they appear to be lured into this bad behaviour. Thus most begin to think of the risks only after starting to smoke and gaining what to them is new information about health risks. Similarly, the detailed calorific information printed on chocolate bars does not seem to convince us much to reduce consumption.

Anchoring, comparative ignorance and ambiguity aversion

Anchoring is the term Tversky and Kahneman (1974) gave to the notion that people prefer to rely on apparently worthless information rather than none to solve problems. For example, when asked about the number of African member states in the United Nations and given the winning number derived from a spinning roulette, many people would rely on this seemingly unrelated piece of information to derive an answer. The motivation behind this odd problem-solving heuristic is the fear or aversion from ambiguity. Ellsberg (1961) argued that people generally prefer to bet on known rather than unknown probabilities. Fox and Tversky (1995) suggest that ambiguity aversion is driven by comparative ignorance, that is, by the comparison with more familiar events or more knowledgeable individuals, and it diminishes or disappears in the absence of such a comparison. Fox and Weber (2002) also assert the importance of the choice context in the degree of ambiguity aversion.

Regret aversion and forgone payoffs

Every so often, we contemplate what might have been if we had chosen differently. Loomes and Sugden (1982) postulate that choice behaviour is affected not only by the attractiveness of a considered alternative, as in EUT and Prospect Theory, but also from the anticipation of regretting not choosing a forgone alternative. Regret Theory predicts an emotion noted as regret aversion, that is, the greater the feeling of regret, the less attractive is the chosen alternative. Accordingly, prior to choosing, the choice-maker compares 'what is' under a particular state with 'what might have been' for an alternative under the same state – which results in anticipated regret or rejoice (the opposite of regret). However, Zeelenberg et al. (1996) argue that experiential information influences the perception of regret. In order to evaluate an alternative by comparing 'what is' with 'what might have been', the choice maker must learn, *ex post*, what is implied by 'what might have been'. Larrick (1993) suggests that it seems reasonable to assume that information about what definitely would have occurred has a greater potential for regret than abstract knowledge of what was statistically likely to occur. Accordingly, Humphrey (2004) postulates that the effect of feedback information (following a choice) on the anticipated emotion of rejoice or regret depends on whether the choice maker can discover what state of the world she currently occupies – one where the forgone alternative would have made her better off or not.

To summarize, here we have discussed different behavioural theories and generalizations that demonstrate how a wide range of seemingly contradictory behaviours can arise from different heuristics and their combinations. In the next sections we will show how these heuristics play an important role in the influence of information on travel behaviour.

2.3 Process Theories of Behaviour

Before going on, it is important to note to the reader that the theories we have so far discussed relate to decision rules involved in human choice. A very different strand of behavioural literature focuses on the effects of information on the process of forming choices through beliefs and attitude updating. The Theory of Planned Behaviour (TPB) (Ajzen, 1991) asserts that behaviour's antecedents are beliefs held about choice options that influence attitudes toward the choice options and generate intention to behave in a certain way. Eventually, the option is chosen that is believed to result in the most favourable behavioural consequences. TPB also states that conformation to social norms plays a role in the decision whether or not to change behaviour, that is, perceived approval of the behaviour by peers. TPB highlights the importance of perceived behavioural control over factors that might hinder or facilitate behaviour, which will positively contribute to behavioural change. Norm Activation Theory (NAT) (Schwartz, 1977; Schwartz and Howard, 1981) emphasizes intrinsic norms and moral convictions about how to behave, mainly in the context of pro-social, altruistic behaviour. Theory of Interpersonal Behaviour (TIB) (Triandis, 1977; 1980) stresses the effect of habit on the formation of intention and behaviour. The more habitual the behaviour, the less conscious are the decisions that are made, and the less likely is it that an intention is generated in response to a change in context. TIB further asserts that an affective component, related to the emotion toward the behaviour's outcome, is also relevant to understand the formation of intention and consequent behaviour.

3 TRAVEL INFORMATION AND TRAVEL BEHAVIOUR

3.1 Typology of Travel Information and Transport Network Environments

Before going into an in-depth review of the literature, it is important to clarify to the reader the three types of information we discuss. Assuming a hypothetical choice between two alternative routes, lacking any other source of information, travellers' choices are usually based on *feedback-based experiential information* generated from knowledge gained from learning of past experiences, for example that on most days route A takes 20 minutes. *Descriptional information* usually consists of information about prevailing conditions such as current or predicted travel times. It can be provided either statically pre-trip, for example via the Internet, or dynamically en-route, for example through a variable message sign or personal nomadic devices like smartphones. An example for descriptional information is that travel time to the destination on route A is 20 minutes, and on route B it is 25 minutes. *Prescriptive information* usually suggests a recommended alternative like the route with the quickest travel time, for example advising the traveller to take route A (rather than B). Such prescriptive information can be provided by traveller planners and sat-nav apps. Traveller response is generally defined here in terms of compliance, that is, acceptance of the system's suggestions. Experiential information is usually *ex post*, that is, after a choice is made, whereas descriptional and prescriptive information are *ex ante*, that is, prior to a choice.

In addition to being categorized according to the types of information available, travel behaviour studies have been exploring the effects of information using two main types of choice environments. The first can be categorized as non-competitive transport networks. These are synthetic environments where each individual traveller considers a decision while the outcome of her decision has no influence on the social or system outcome. It is common that these types of networks are constructed considering two alternatives, for example routes or modes, for simplicity. Though far from reality where choices interact and have collective consequences such as traffic congestion, these settings are useful for testing different behavioural hypotheses under controlled conditions, often using a travel simulator apparatus. This is basically a computerized simulation which can account for different levels of realism or complexity and can include different kinds of travel information for hypothesis testing. Other methods include the use of numerical simulations based on synthetic data for postulating models and understanding potential dynamics. We rarely find field studies in this kind of setting.

Unlike non-competitive networks, in competitive networks, individual choices aggregate to form flows on the network and these flows of travellers result in a social outcome in terms of travel time distributions experienced simultaneously by all travellers using the network in that time period. Competitive networks experience congestion directly, whereas in non-competitive networks, the traffic conditions are externally manipulated through exogenous travel time distributions. Traffic simulations are the common way to study network dynamics. However, some examples include field studies or natural experiments involving a network manipulation like a road closure, and behaviour is tracked throughout a certain designated period of time. In competitive networks, equilibrium (a situation in which all travellers will choose the route which requires the least travel time or cost for them) is used as a criterion for stability. However, with the presence of information, strange patterns of behaviours can emerge, and the network can be categorized by complexity as its own (aggregative) behaviour is very different from the behaviour of its individual users. Often, the rate of exposure to information or information penetration is important to predict how the network will behave. Moreover, with competition, information can be used to shape behaviours toward social learning and possible cooperation, leading to development of new traffic management regimes.

To make the discussion easily understandable, we will first review the studies conducted in non-competitive networks for the different types of information, followed by a review of the studies applied in competitive networks. We aim at reviewing mostly recent studies from the past ten years or so. The reader interested in literature which precedes this will find an excellent review by Chorus et al. (2006a).

3.2 Travel Behaviour with Information in Non-competitive Networks

Travel behaviour with experiential information

Lacking other sources, travellers will base their choices on information based on retrospective memory of the travel times experienced in previous trips. Feedback-based experiential information reinforces learning but, at the same time, it is also important to remember that the availability of such information is a function of sampling what is stored in long-term memory. Over time learning is internalized, forming the basis for habitual travel behaviour (Gärling and Axhausen, 2003). Various studies applied a day-to-day learning approach based on the rational choice utilitarian framework. For example, Horowitz (1984) used the notion of utility updating over time. Cascetta and Cantarella (1991) used a similar approach to model choices in a stochastic dynamic traffic assignment. Positive studies based on empirical observation of behaviour in the lab have also been conducted to gather knowledge on learning effects. Avineri and Prashker (2003; 2005) showed the workings of the payoff variability effect whereby a route became more attractive when its travel time variability was increased, moving the choice shares towards randomness. Ben-Elia et al. (2008) corroborated these findings, but showed explicitly that compared to a network where travel time variability is altogether low, increasing this variability results in choices closer to random. Ben-Elia and Shiftan (2010) explained that this behaviour can be asserted as a result of the recency effect and decaying memory whereby the effect of recent outcomes outweighs outmoded ones. Thus alternatives positively associated with recent outcomes will be likely to be repeated, creating a classic learning curve. Bogers et al. (2007) showed that respondents who were provided with forgone payoffs (that is, feedback on chosen and non-chosen alternatives alike) performed better in terms of cost minimization, though these benefits decreased over time as more experience was accumulated.

Travel behaviour with descriptional information

Ex-ante exposure to pre-trip or en-route information changes the traveller choice problem to one based on descriptional information. Most studies conducted on descriptional information effects adopt a non-repeated one-shot choice situation without feedback (that is, the effect of experiential information is neutralized). In this domain, the dominant behavioural paradigm is utilitarian (for example Arentze and Timmermans, 2005; De Palma and Picard, 2005). Researchers commonly postulate a model to explain and predict behaviour and then either empirical or synthetic data is used to corroborate or refute the postulated model. Ettema and Timmermans (2006) provided a classical example; they showed using numerical simulations how the negative effects of travel time uncertainty can be reduced by providing travellers with travel time information, which improved their estimate of the expected travel time, thereby reducing scheduling costs. The main caveat to this approach is that this postulated model cannot be empirically verified without real data. Chorus et al. (2013) demonstrated a discrete-choice rational expectation model for information acquisition and perception. They postulated that travellers acquire information based on the utility of the anticipated travel choice situation after the information is received. Once information has been acquired, the received messages alter the travellers' perceptions of the number of known travel alternatives or the variability attached to their attributes. Subsequently, the traveller must again choose from a choice set containing options for additional information acquisition as well as travel alternatives. Model estimates were based on travel simulator data and revealed that people adhere to myopic forward-looking behaviour in the joint task for information acquisition and travel choice based on it.

Another issue is that travellers may well behave strategically by planning ahead for information provided en route, acquiring information as long as there is a reasonable gain from it (Bonsall, 2004). In this sense, the fact that a link downstream has an installed VMS can make it more attractive, as demonstrated by Razo and Gao (2011) in a lab experiment. They applied a rank-dependent utility model for explaining the choice among routing policies where the route attributes are dependent on the occurrence of stochastic incidents leading the traveller to consider not only the route but also the probability that an incident may occur on the way. As only by passing through the VMS will the knowledge be acquired of which state of the world is likely to happen, the VMS-installed link becomes more attractive and strategic behaviour arises.

Framing and Loss aversion have also been tested in travel behaviour studies where Prospect Theory is applied and found to have added explanatory value. Katsikopoulos et al. (2002) showed how a more risky route can be preferred when the range of travel times is presented. They demonstrated in a static choice experiment that a route that is perceived as a relative expected loss is preferred when its travel time range is greater than the reference, indirectly implying risk seeking. Avineri and Prashker (2004) and Avineri and Bovy (2008) showed how framing effects and loss aversion in one-shot situations are evident when prospects are described in travel time terms (in contrast to money in the original theory). Avineri and Waygood (2013) demonstrated loss aversion and framing effects when travel alternatives were described in terms of CO_2 emissions. In a synthetic static set-up, Gao et al. (2010) showed that strategic planning ahead is also consistent with Prospect Theory. Despite providing more realism to the choice process, a main caveat of Prospect-Theory-based studies is the arbitrary selection of the perceived reference point which poses a considerable challenge for modellers – since it is not well defined in the literature. Moreover, this reference point is not necessarily stationary but possibly dynamic, changing from one context to the other, and over time implying that information provided in different stages of the decision process, for example pre-trip or en route, may well have different behavioural outcomes.

Regret Theory has been suggested as an alternative description-based model to both Prospect Theory and EUT. Chorus et al. (2008) developed a model based on regret minimization rather than utility maximization. The Random Regret-Minimization (RRM) model provides a straightforward and intuitive way to incorporate the notion that travellers, when faced with knowledge limitations, may wish to postpone their choice and search for more information first. The developed model was estimated on data from a multimodal travel simulator, where participants could choose between travel alternatives with risky travel times and costs, and the option of travel information acquisition. Semi-compensatory decision-making and choice set composition effects like the compromise effect, emerge as RRM model features. Chorus et al. (2006b) integrated Bayesian updating into a regret-based framework of travel choice. Numerical simulations of the model provide insights into the mechanisms behind information use and effect in a mode-choice context where a traveller chooses between car and public transport but has an overall preference for car travel. These simulations show that the perceived value of acquiring public transport information is limited by a number of factors. Furthermore, even in the case where public transport information is acquired, and the message is favourable to public transport, its impact on mode choices will also remain limited because of the effect of regret aversion.

Travel behaviour with both experiential and descriptional information

A main drawback of description-based studies is that although the traveller receives information regarding the travel alternative routes or modes, there is no ability to learn from this information. This means that the long-term effects of information are ignored and the relationships between descriptional and experiential information are not resolved. However, in real travel conditions, it is more than likely that a traveller will use the information to make choices and the outcomes of these choices will in turn influence subsequent choices. The extent to which descriptional information will influence future behaviour depends to a large degree on how it is received, that is, whether it is general pre-trip or travel-specific provided en route. Two main issues emerge in the literature: the formation of risk attitudes in response to information and the perception of reliability in the provided descriptional information.

Abdel-Aty et al. (1997) asserted in lab experiments that travellers will tend to exhibit risk aversion when faced with pre-trip travel time information. Avineri and Prashker (2006) compared the effect of providing respondents a-priori static pre-trip information describing average expected travel times for two routes in addition to receiving feedback information about the chosen alternative. They demonstrated that informed respondents were more risk averse and preferred a more reliable route compared to a control group that could only learn through experiential information. Conversely, Ben-Elia et al. (2008) provided respondents with dynamic en-route information describing the ranges of travel times. They show that compared to a control group which received only experiential information but no descriptional information, respondents provided with descriptional information learn much faster. In addition, they found risk-seeking tendencies (that is, preferring a shorter and riskier route) in the short run. This tendency dissipates in the long run as experiential information becomes more dominant. The control group, in contrast, showed mostly risk-averse behaviour. Informed respondents also tended to switch routes more often while the average payoffs (network travel times) between the two groups were mostly the same. Ben-Elia and Shiftan (2010) explained that this behavioural difference arises from different mental processes associated with formation of long-term memory. Whereas the control group had only experience to rely on leading to a recency effect, descriptional information allowed the treatment group to generate a mental anchor on which to build experiential information more efficiently, allowing them to better sample from long-term memory.

Another finding in combined studies is the apparent weakening of framing and loss aversion in repeated travel choices as found by Erev et al. (2008) for monetary payoffs. Ben-Elia and Shiftan (2010) did not find behaviour to be completely consistent with Prospect Theory. Although differences relative to a reference point (mean travel time in this case was arbitrarily chosen) seem to have some significance in explaining route-choice behaviour, a real difference was neither identified between gains and losses nor evidence for loss aversion detected. However, given that only one reference point definition was tested based on average expected travel time, it is difficult to generalize from their findings on the appropriateness of Prospect Theory in dynamic travel contexts.

Combining both descriptional and experiential information also influences the sense of regret aversion. Ben-Elia et al. (2013a) revealed that provision of descriptional information increases the sensation of regret aversion. In contrast, with only experiential information regret is much lower. They explained this anomaly by the fact that descriptional information allows travellers to learn better the state of the world that has occurred, which will result in more regret aversion, whereas the control group which only has experiential information has less ability to solve this, resulting in less regret. An important contribution of this work is that the source of regret aversion is actually the experiential information, that is, the feedback received following the choice as asserted by feedback-conditioned regret theory; however, its effect is amplified by the provision of descriptional information.

Affective influences on travel behaviour are particularly accentuated when combining both types of information in a mode choice context. In a lab experiment, Innocenti et al. (2013) studied the effect of information on the choice between car and public transport. They found a marked preference for cars, with a strong recency effect leading to mode stickiness. The strength of affective dominance was apparent even when the car performed significantly less well than the public transport alternative. This shows that affective dominance can cancel out the positive benefits of information for encouraging mode changes. As pointed out by Steg (2005), cars are perceived as symbols of status and freedom, with the result that costs are often underestimated and a strong label effect for the car mode is often found. With similar conclusions, Farag and Lyons (2010; 2012) asserted in a field-based study that the desire to consult public transport information is affected by attitudes, subjective norms and past behaviour (consistent with TPB and TIB). They found that consulting information is influenced by the propensity to consider using public transport rather than information influencing mode choice. Furthermore, past experience and travel attitudes had the strongest effect on the use of public transport information to compare car transport with public transport. Respondents tended to travel more often by car than by public transport after comparing travel modes.

The combination of information sources raises a further important issue, which is the influence of the perceived reliability or accuracy of descriptional information. Ben-Elia et al. (2013b) asserted that behavioural differences attributed to pre-trip and en-route information can be explained by perceptions of information reliability. In Ben-Elia et al. (2008), the actual travel times that respondents experienced were always drawn from within the descriptional travel time range, thus no dissonance will have been perceived between the descriptional information provided and the experiential information following the choice. This could lead to greater confidence and promote respondents to take more risks initially until learning sets in. In contrast, larger discrepancies between the descriptional pre-trip average and the experiential choice outcome as in Avineri and Prashker (2006) could lead to lack of confidence in the received information and greater risk aversion as a consequence. The two aforementioned studies illustrate the possible associations that could exist between the perceived trust in the reliability of information on one hand and changes in travellers' risk attitudes on the other hand.

Travel behaviour with prescriptive information

The key issue in studying behaviour with prescriptive information is the degree of compliance with the suggestions. Chorus et al. (2009), using a Bayesian approach (that is, travel time probabilities are conditional on descriptional information provided) and numerical simulations showed that compliance is negatively associated with the perceived unreliability of the information and the uncertainty regarding the travel time differences between (two) alternative routes. In a lab experiment which combined all three types of information and a tri-route network which included risky, reliable, and useless route alternatives, Ben-Elia et al. (2013b) asserted that prescriptive information has the largest behavioural impact, followed by descriptive and experiential feedback information. However, decreasing the accuracy of prescriptive information shifts choices not only from the riskier to the reliable route but also to the useless alternative. This finding seems to be in line with the comparative ignorance postulation and the aversion of ambiguity. Thus, travellers will prefer to rely on information even if they sense that it is not very helpful, and only to avoid being left completely in the dark.

3.3 Travel Behaviour and Effects of Information in Competitive Networks

Traffic simulations, in particular assignment models, have been used for many years to understand network dynamics (Daganzo and Sheffi, 1977; Daganzo, 1983). One of the key assumptions applied in these models is that the agents have perfect information on all the travel alternatives for each pair of origins and destinations. Here, uncertainty is accommodated through stochasticity, that is, that the outcomes of the choices may have indirect uncertainty. This can be attributed to non-recurring incidents which reduce capacity or variability in the demand. However, once information effects need to be accommodated directly, the assumption of perfect information needs to be abandoned. Another issue is that of network stability. Traffic simulation models, both static and dynamic, work on the principle of convergence towards stable user equilibrium (Arnott et al., 1991; 1996; 1999). User Equilibrium (Wardrop, 1952; Beckmann et al., 1956) is a condition whereby behaviours do not change because there is no possibility to gain an advantage to incentivize this change. It is also a suboptimal state, as it does not reduce the overall time or cost in the system. Although a more optimal state would be a social optimum where total travel time is minimized, this is at present regarded as impossible to achieve, given the contradiction with the second important assumption adopted in network analysis: competition. Competition is a direct extension of the rational choice assumption, in that agents are inherently selfish and so they compete over the available capacity, trying to travel the quickest way (that is, maximizing individual payoffs) with the end result that inefficient equilibrium is attained. Traveller competition also determines travel times and their distributions on the networks in an endogenous manner.

There have been several attempts to further apply Prospect Theory (PT) to improve network equilibrium models, substitute for decision rules, and examine the effect of the reference point value on such equilibrium (for example Avineri, 2006; Connors and Sumalee, 2009; Tian et al., 2012). An uncertainty effect (UE) concept based on regret aversion was also developed by Chorus (2010). However, most of this work in this field is theoretical by nature and the functional forms of PT apply parameter values that were obtained through experiments in general, non-transport contexts (such as Tversky and Kahneman, 1992).

Behavioural studies in simulated and real networks found a more complex view than what appears from traffic models. These studies were exclusively about car traffic (route choice) and involved either experiential information or a combination of experience and description. Network studies with prescriptive information are still very few and are centred on compliance with descriptional information in mostly synthetic settings in the context of traffic optimization (for example Leontiadis et al., 2011; Dia and Panwai, 2010). Exceptional are Helbing et al. (2002), who develop an analytical dynamic framework based on fractional provision of prescriptive information to switch routes. They demonstrate how such a framework can bring the network closer to user equilibrium in an agent-based model.

In relation to experiential information, Nakayama et al. (1999; 2000; 2001) studied the dynamics of behaviour in a simulated network with bounded rationality. They show that reinforced learning that allows agents to adapt to uncertain travel times inhibits convergence to equilibrium. They also showed a greater degree of heterogeneity in behaviours compared to what was previously considered, including situations characterized by delusions and freezing of route choices. Selten et al. (2007) found similar results in an experiment with human participants. The main findings from learning-based studies are that network traffic is often more volatile compared to assignment model predictions and equilibrium is not always attained as models assume. Moreover, as shown by Rapoport et al. (2009), these fluctuations can also result in large inefficiencies when the network's typology results in a Braess' paradox where a useless alternative will cause overall energy spent on the network to grow rather than to decrease. Unlike simulations, where the paradox is seen as an unstable condition, human respondents fail to learn that this alternative only results in greater loss for all travellers.

Effects of descriptional information were also studied in competitive networks. Mahmassani and Liu (1999) conducted competitive experiments in a three-route network with connectors between the routes. They found that pre-trip or en-route real-time information on prevailing traffic conditions increases the route switches at the point of information provision. The reliability of information influences commuters' pre-trip departure time and route switching decisions, as well as en-route path switching decisions. Similarly to Ben-Elia et al. (2008), they found information availability tends to induce greater frequency of route switching, both pre-trip and en route. Furthermore, travellers become more prone to switch routes when they perceive late arrival by following the current path than when they perceive early arrival. This is further evidence of risk-seeking behaviour in the face of a potential loss. In a field study Fujii and Kitamura (2000) investigated the effect of descriptional information in a real-life road closure situation. They found that drivers were more sensitive to descriptive travel time information than to experiential information. This result could be magnified by the high risk of delays involved in the road closure in the short run. This result suggests behaviour found in non-competitive networks (for example, Ben-Elia et al., 2008) still holds also for competitive ones. Lu et al. (2011) introduced en-route real-time information on the occurrence of a non-recurring incident and feedback information on non-chosen routes in a competitive experiment. En-route real-time information reduces overall network travel time and reliability.

The effects are similar to those found in non-competitive networks, that is, that information expedites learning. However, too much information might result in less efficiency and time loss. Extra feedback on forgone alternatives brought the network closer to a user equilibrium outcome. These results may be attributed to induced regret. Selten et al. (2007) conducted an experiment with human subjects mimicking a live network. They compared the behaviour with feedback information on a chosen route with a condition where information was also given on the performance of the forgone route. Similarly to Lu et al. (2011), less switching was observed with forgone feedback and the network was closer to a stable equilibrium compared to the condition where respondents could only receive experiential information. It seems that when more complete information is provided, travellers also simultaneously learn which routes are preferable, which paradoxically expedites the convergence of traffic on the road network towards the user equilibrium – as predicted by the same models that assumed perfect information from the start. Lu et al. (forthcoming) investigated strategic behaviours in a competitive network with human respondents and real-time incident information provided on a branch of one of the main routes. They found that the network was overall more efficient with information. However, respondents did not reduce switching substantially; instead they replaced switching at the origin to switching route at the branch where information was provided. This result suggests that travellers may well behave strategically, but that links where information is provided will become saturated. An important result from this study is that equilibrium was never retained at any condition, that is, information may increase efficiency but it does not necessarily bring greater stability to the network. Moreover, information does not bring the network towards a social optimum. A note of caution is that results obtained from network-based studies are prone to depend on the specific network topology and design capacities.

Ben-Akiva et al. (1991) and Mahmassani and Jayakrishnan (1991) suggested that information adversely affects traffic patterns in three ways: concentration, overreaction and oversaturation. Concentration occurs when travellers make more similar choices that exacerbate congestion. Concentration is an equilibrium phenomenon in the sense that individual travellers do not regret the choices they make. Overreaction occurs when travellers fail to anticipate how other travellers will react, and collectively respond too much to new information. Overreaction is out-of-equilibrium behaviour since those who respond end up worse off. Oversaturation occurs if drivers are faced with too much information, and either fail to use it effectively or become overwhelmed and resort to heuristic decision rules. The empirical and simulated studies reviewed so far provide an indication that adverse effects are not entirely hypothetical. Emmerink et al. (1998), who analysed two-route networks with linear link travel cost functions, and Emmerink et al. (1995a; 1995b) concluded that beyond a penetration rate of about 20 percent, information has negative effects due to concentration and overreaction. The critical penetration rate can depend on the type of information provided and the degree of inertia in driver response. These inefficiencies have been found to occur in simulated networks with both recurring and non-recurring congestion effects.

3.4 The Role of Information in Shaping Travel Behaviours

Thus far, the discussion on competitive networks has focused on the possible benefits and disbenefits of descriptional travel information in a dynamic context of day-to-day traffic changes. This information supply strategy does not bring the network closer to an optimal state known as the social (or system) optimum. In fact, the information only allows travellers to maximize (minimize) their individual payoffs (costs/travel times) which maintains the inherent inefficiencies of the transport network. Although often useful in helping travellers avoid non-recurring incidents and bottlenecks, it seems information provides travellers mostly with an illusionary sense of self-control over ambiguous travel conditions (Kemel, 2013). Nevertheless, some ideas for changing this are appearing in the literature. First, the assumption that humans are necessarily selfish is seen in contradiction to studies that show that sometimes people are willing to behave empathetically towards each other and cooperate (Kim, 2011). In this sense, information can be used to shape and motivate good behaviour. Bartle et al. (2013) describe how travellers are willing to freely exchange information on traffic conditions through social media without apparent extrinsic rewards. This behaviour is also evident in the use of social navigation apps like Waze, where travellers can report incidents to the rest of the community of travellers and this information is commonly shared. Experiments conducted by Helbing et al. (2005) demonstrate how, cooperation emerges, when travellers learn to take turns on alternate routes, resulting in increased energy efficiency. Helbing (2004) also showed that information that represents the target for a social optimum can guide travellers to cooperate on the use of the network's capacity. Sunitiyoso et al. (2011) presented an idea based on conformity to a group target as a possible idea to utilize social interaction as a way to motivate a mode change. Following Olson (1965), who suggested smaller groups are more likely to reach consensus and avoid pluralistic ignorance, they showed experimentally that using simple communication tools such as texting small groups can help people learn to cooperate on the use of a public good like transport – in this case collective decisions on the use of a bus alternative rather than the car. These are emerging ideas, but they elucidate the potential of ICT to move to a new, more proactive role in managing traffic and transport rather than just providing information. This presents planners and designers of information systems with new challenges – how to encourage, through the design of content and context of travel information, pro-social and pro-environmental travel behaviours and eliminate the external costs associated with travel choices (for example CO_2 emissions).

Thaler and Sunstein (2008) advocated the use of 'Choice Architecture' to influence behavioural change; they illustrate how 'nudges', small features designed in the environment of choice making, could help individuals to overcome cognitive biases, and to highlight the better choices for them and increase the effect of behavioural change – without restricting their freedom of choice, and without making big changes to the physical environment, the set of choices, or the economic attributes of the choices (Avineri, 2012). Choice architecture might be a useful approach to design information content that facilitates the emergence of cooperation and optimal self-organization to deal with congestion efficiently without resorting to penal measures, like pricing and tolls. It could even possibly provide optimal solutions in the case of network disruptions and emergency evacuation where, without cooperation, traffic will run into utter chaos. For example, loss framing can be incorporated in the design of a variety of information-based measures to promote travel behaviour change (Avineri and Waygood, 2013). Framing is only one example of how contextual design of information might be effective in influencing human behaviour; other examples of choice architecture include the design of smart *defaults* (people tend to 'go with the flow' of pre-set options), *priming* (people's behaviour may be altered if they are first exposed to certain sights, words or sensations) and incorporating information on *norms* (people are influenced by what others do or expect them to do) (for a review, see Avineri, 2012).

In this respect, travel information can be used in personalized campaigns and travel plans that work on the provision of information to change attitudes and beliefs about certain travel alternatives (usually in order to move travel to more sustainable modes). Often noted as soft policies (to distinguish from hard policies like road pricing), these interventions have become quite popular in recent years. A non-exclusive list of international implementation examples include: Australia (Zhang et al., 2013), Japan (Fujii et al., 2009; Fujii and Taniguchi, 2006), UK (Bamberg and Möser, 2007), and Sweden (Friman et al., 2013). The interested reader can find several more examples of this approach in a special issue coedited by the first author on motivating sustainable travel behaviours (Ben-Elia and Shiftan, 2013).

4 CONCLUSIONS: THREATS AND OPPORTUNITIES OF TRAVEL INFORMATION AND FUTURE OUTLOOK

In this chapter we have tried to plot the rather large amount of literature on the behavioural effects of travel information. Understanding how information influences travel behaviour requires adding a question mark to the fundamental behavioural assumptions used in transport science and practices. Rational choice and utilitarianism continue to be the dominant normative paradigm in transport. Behavioural scientists have suggested that human cognition and behaviour are better described by bounded rationality and the application of heuristics. Some of the leading alternative theories such as reinforced learning, framing and loss aversion, affect, regret and ambiguity have been discussed. The review of the transport literature suggests that effects of information on travel behaviour are determined both by the type of information and the choice environment. The literature also suggests a hierarchy of information influence, with prescriptive information at the top, followed by descriptional and experiential information.

With this in mind, what are the main opportunities and threats of travel information? Clearly, the common claim that information is always a good thing for the individual and the system is both naive and potentially dangerous. In terms of the opportunities, most studies agree that descriptional information is useful and beneficial from an individual's point of view, as it assists in speeding up learning of travel time distributions under unfamiliar and ambiguous travel conditions. This is especially relevant, given that learning is inhibited by the level of uncertainty in the choice environment, as noted by the payoff variability effect. Information is likely to reduce stress, increase a sense of self-control and would be likely to be positively associated with subjective well-being. Moreover, descriptional information works in combination with experiential information and learning and seems to be effective mainly in the short run, whereas in the long run, travellers tend to make more habitual choices once learning is internalized. However, information is also useful from a network perspective, particularly to allow travellers to overcome the disruptions associated with non-recurring incidents like accidents or bad weather, and avoid excessive delays.

Notwithstanding, the literature describes serious and potential threats of the current affairs in provision of travel information. Real-time information in the form of expected travel times supplied dynamically seems to increase travellers' tendency to switch routes more often. This behaviour is likely to be associated with risk seeking in the face of losses due to potential delays. Descriptional information also induces strategic travel behaviour, with downstream links that resolve uncertainty becoming more attractive and promoting greater switching en route. Possibly such strategic behaviour will become less prominent once the supply of information is independent of physical location, for example incident information provided remotely to smartphones. At the network level, descriptional information does not seem to contribute to greater stability or to attaining equilibrium. Moreover, even if information is provided, attitudes and affective stickiness seem to inhibit both acquisition of information and change of behaviour, for example, to a different mode. Too much information is also problematic. Provision of information on the performance of forgone alternatives seems to generate excessive regret aversion and results in the network reaching traffic conditions resembling user equilibrium which is inefficient for the greater good. Reliability of information is also an issue of concern. As shown, compliance rates are not reduced by unreliable and inaccurate information. Consequently, travellers will choose even useless alternatives. Comparative ignorance and ambiguity aversion seem to be the main culprits of this anomaly. The problem is that information reliability is closely associated with traffic conditions. Incidents and disruptions are both hard to predict, and providing accurate information once they occur is difficult to manage. Therefore, sometimes less information might be a better solution networkwise. A further concern, based on evidence of strategic behaviour, is that proliferation of information will gradually bring traffic conditions to resemble traffic simulations predicted by perfect information, resulting in greater network inefficiencies and energy consumption. This real risk was demonstrated both for recurring (known) congestion and non-recurring (unknown) congestion conditions.

The main caveat of current information provision strategies is failure in making substantial changes to the competitive nature of the transport system and its use. Looking forward into the future, we can say that technological developments could possibly create the conditions to change this substantially. Open communication channels between travellers, vehicles and the infrastructure provide new opportunities to optimize mobility through cooperation and social learning. As noted, both common road knowledge examples and empirical studies suggest that travellers are willing to behave more empathetically and prosocially towards each other. Cooperative behaviour has so far been overlooked in transport modelling. However, pervasive ICT, the Cloud and crowdsourcing could well be the basis for applying choice architecture techniques to design information that motivates individual travellers to cooperate and use the available infrastructure more efficiently and sustainably, possibly reaching social optima. Cooperative mobility can be embedded gradually with the deployment of fully autonomous vehicles, where the traveller is not preoccupied with the motoric task of driving the vehicle. Furthermore, evacuation strategies put in place to manage traffic safety in the face of natural disasters and extreme weather events (now likely to be more frequent in the face of global warming) that cause severe network disruptions and infrastructure failures will succeed only if travellers cooperate and comply with socially-optimal guidance information.

ACKNOWLEDGEMENTS

The authors would like to kindly thank three anonymous reviewers and the editors for their helpful and insightful comments.

REFERENCES

Abdel-Aty, M., Kitamura, R. and Jovanis, P. (1997), Using stated preferences data for studying the effect of advanced traffic information on drivers' route choice, *Transportation Research Part C: Emerging Technologies*, **5**(1), 39–50.

Ajzen, I. (1991), The theory of planned behavior, *Organizational Behavior and Human Decision Processes*, **50**, 179–211.

Arentze, T.A. and Timmermans, H.J.P. (2005), Information gain, novelty seeking and travel: a model of dynamic activity-travel behavior under conditions of uncertainty, *Transportation Research Part A: Policy and Practice*, **39**, 125–45.

Arnott, R., de Palma, A. and Lindsey, R. (1991), Does providing information to drivers reduce traffic congestion? *Transportation Research Part A: Policy and Practice*, **25**, 309–18.

Arnott, R., de Palma, A. and Lindsey, R. (1996), Information and usage of free-access congestible facilities, *International Economic Review*, **37**, 181–203.

Arnott, R., de Palma, A. and Lindsey, R. (1999), Information and time-of-usage decisions in the bottleneck model with stochastic capacity and demand, *European Economic Review*, **43**, 525–48.

Avineri, E. (2006), The effect of reference point on stochastic network equilibrium, *Transportation Science*, **40**(4), 409–20.

Avineri, E. (2012), On the use and potential of behavioural economics from the perspective of transport and climate change, *Journal of Transport Geography*, **24**, 512–21.

Avineri, E. and Bovy, P.H.L. (2008), Identification of parameters for a prospect theory model for travel choice analysis, *Transportation Research Record*, **2082**, 141–7.

Avineri, E. and Prashker, J.N. (2003), Sensitivity to uncertainty: need for a paradigm shift, *Transportation Research Record*, **1854**, 90–98.

Avineri, E. and Prashker, J.N. (2004), Violations of expected utility theory in route-choice stated preferences: certainty effect and inflation of small probabilities, *Transportation Research Record*, **1894**, 222–9.

Avineri, E. and Prashker, J.N. (2005), Sensitivity to travel time variability: travelers' learning perspective, *Transportation Research Part C: Emerging Technologies*, **13**, 157–83.

Avineri, E. and Prashker, J.N. (2006), The impact of travel time information on travelers' learning under uncertainty, *Transportation*, **33**, 393–408.

Avineri, E. and Waygood, E.O.D. (2013), Applying valence framing to enhance the effect of information on transport-related carbon dioxide emissions, *Transportation Research Part A: Policy and Practice*, **48**, 31–8.

Bamberg, S. and Möser, G. (2007), Why are work travel plans effective? Comparing conclusions from narrative and meta-analytical research synthesis, *Transportation*, **34**, 647–66.

Barron, G. and Erev, I. (2003), Small feedback-based decisions and their limited correspondence to description-based decisions, *Journal of Behavioral Decision Making*, **16**, 215–33.

Bartle, C., Avineri, E. and Chatterjee, K. (2013), Online information-sharing: a qualitative analysis of community, trust and social influence amongst commuter cyclists in the UK, *Transportation Research Part F: Traffic Psychology and Behaviour*, **16**, 60–72.

Beckmann, M.J., McGuire, C.B. and Winsten, C.B. (1956), *Studies in the Economics of Transportation*, New Haven, CT: Yale University Press.

Ben-Akiva, M. and Lerman, S.R. (1985), *Discrete Choice Analysis: Theory and Application in Travel Demand*, Cambridge, MA: MIT Press.

Ben-Akiva, M., de Palma, A. and Kaysi, I. (1991), Dynamic network models and driver information systems, *Transportation Research Part A: Policy and Practice*, **25**, 251–66.

Ben-Elia, E. and Shiftan, Y. (2010), Which road do I take? A learning-based model of route-choice behavior with real-time information, *Transportation Research Part A: Policy and Practice*, **44**, 249–64.

Ben-Elia, E. and Shiftan, Y. (2013), Editorial: Understanding behavioural change: an international perspective on sustainable travel behaviours and their motivations. Selected papers from the 12th World Conference on Transportation Research, *Transport Policy*, **26**, 1–3.

Ben-Elia, E., Erev, I. and Shiftan, Y. (2008), The combined effect of information and experience on drivers' route-choice behavior, *Transportation*, **35**, 165–277.

Ben-Elia, E., Ishaq, R. and Shiftan, Y. (2013a), 'If only I had taken the other road. . .': Regret, risk and reinforced learning in informed route-choice, *Transportation*, **40**, 1–25.

Ben-Elia, E., Di Pace, R., Bifulco, G.N. and Shiftan, Y. (2013b), The impact of travel information's accuracy on route-choice, *Transportation Research Part C: Emerging Technologies*, **26**, 146–59.

Bernoulli, D. (1738), Exposition of a new theory on the measurement of risk, Comentarii Academiae Scientiarum Imperialis Petropolitanae, translated and reprinted in 1954, *Econometrica*, **22**, 22–36.

Bogers, E., Bierlaire, M. and Hoogendoorn, S. (2007), Modeling learning in route choice, *Transportation Research Record*, **2014**, 1–8.

Bonsall, P. (2004), Traveller behavior: decision-making in an unpredictable world, *Journal of Intelligent Transportation Systems*, **8**, 45–60.
Camilleri, A.R. and Newell, B.R. (2011), When and why rare events are underweighted: a direct comparison of the sampling, partial feedback, full feedback and description choice paradigms, *Psychonomic Bulletin & Review*, **18**, 377–84.
Cascetta, E. and Cantarella, G.E. (1991), A day-to-day and within-day dynamic stochastic assignment model, *Transportation Research Part A*, **25**(5), 277–91.
Chorus, C.G. (2010), Regret theory based route choices and traffic equilibria, *Transportmetrica*, **8**, 291–305.
Chorus, C.G., Arentze, T.A. and Timmermans, H.J.P. (2008), A random regret-minimization model of travel choice, *Transportation Research Part B: Methodological*, **42**, 1–18.
Chorus, C.G., Arentze, T.A. and Timmermans, H.J.P. (2009), Traveler compliance with advice: a Bayesian utilitarian perspective, *Transportation Research Part E*, **45**, 486–500.
Chorus, C.G., Molin, E.J. and Van Wee, B. (2006a), Use and effects of Advanced Traveller Information Services (ATIS): a review of the literature, *Transport Reviews*, **26**, 127–49.
Chorus, C., Molin, E., Van Wee, B., Arentze, T. and Timmermans, H. (2006b), Responses to transit information among car-drivers: regret-based models and simulations, *Transportation Planning and Technology*, **29**, 249–71.
Chorus, C.G., Walker, J.L. and Ben-Akiva, M. (2013), A joint model of travel information acquisition and response to received messages, *Transportation Research Part C: Emerging Technologies*, **26**, 61–77.
Connors, R.D. and Sumalee, A. (2009), A network equilibrium model with travellers' perception of stochastic travel times, *Transportation Research Part B: Methodological*, **43**(6), 614–24.
Daganzo, C. (1983), Stochastic network equilibrium with multiple vehicle types and asymmetric, indefinite link cost Jacobians, *Transportation Science*, **17**, 282–300.
Daganzo, C. and Sheffi, Y. (1977), On stochastic models of traffic assignment, *Transportation Research Part B: Methodological*, **14**, 243–55.
Denrell, J. (2007), Adaptive learning and risk taking, *Psychological Review*, **114**, 177–87.
Denrell, J. and March, J.G. (2001), Adaptation as information restriction: the hot stove effect, *Organization Science*, **12**, 523–38.
De Palma, A. and Picard, N. (2005), Route choice decision under travel time uncertainty, *Transportation Research Part A: Policy and Practice*, **39**, 295–324.
Dia, H. and Panwai, S. (2010). Evaluation of discrete choice and neural network approaches for modelling driver compliance with traffic information, *Transportmetrica*, **6**(4), 249–70.
Ellsberg, D. (1961), Risk, ambiguity, and the savage axioms, *Quarterly Journal of Economics*, **75**, 643–69.
Emmerink, R.H.M., Axhausen, K.W., Nijkamp, P. and Rietveld, P. (1995a), Effects of information in road transport networks with recurrent congestion, *Transportation*, **22**, 21–53.
Emmerink, R.H.M., Axhausen, K.W., Nijkamp, P. and Rietveld, P. (1995b), The potential of information provision in a simulated road transport network with non-recurrent congestion, *Transportation Research Part C: Emerging Technologies*, **3**, 293–310.

Emmerink, R.H.M., Verhoef, E.T., Nijkamp, P. and Rietveld, P. (1998), Information policy in road transport with elastic demand: some welfare economic considerations, *European Economic Review*, **42**, 71–95.

Erev, I. and Barron, G. (2005), On adaptation, maximization and reinforcement learning among cognitive strategies, *Psychological Review*, **112**, 912–31.

Erev, I., Ert, E. and Yechiam, E. (2008), Loss aversion, diminishing sensitivity, and the effect of experience on repeated decisions, *Journal of Behavioral Decision Making*, **21**, 575–97.

Ettema, D. and Timmermans, H. (2006), Costs of travel time uncertainty and benefits of travel time information: conceptual model and numerical examples, *Transportation Research Part C: Emerging Technologies*, **14**, 335–50.

Farag, S. and Lyons, G. (2010), Explaining public transport information use when a car is available: attitude theory empirically investigated, *Transportation*, **37**, 897–913.

Farag, S. and Lyons, G. (2012), To use or not to use? An empirical study of pre-trip public transport information for business and leisure trips and comparison with car travel, *Transport Policy*, **20**, 82–92.

Fox, C.R. and Tversky, A. (1995), Ambiguity aversion and comparative ignorance, *Quarterly Journal of Economics*, **110**, 585–603.

Fox, C.R. and Weber, M. (2002), Ambiguity aversion, comparative ignorance, and decision context, *Organizational Behavior and Human Decision Processes*, **88**, 476–98.

Friman, M., Larhult, L. and Gärling, T. (2013), An analysis of soft transport policy measures implemented in Sweden to reduce private car use, *Transportation*, **40**, 109–29.

Fujii, S. and Kitamura, R. (2000), Anticipated travel time, information acquisition, and actual experience: Hanshin expressway route closure, Osaka-Sakai, Japan, *Transportation Research Record*, **1725**, 79–85.

Fujii, S. and Taniguchi, A. (2006), Determinants of the effectiveness of travel feedback programs – a review of communicative mobility management measures for changing travel behaviour in Japan, *Transport Policy*, **13**, 339–48.

Fujii, S., Bamberg, S., Friman, M. and Gärling, T. (2009), Are effects of travel feedback programs correctly assessed? *Transportmetrica*, **5**(1), 43–57.

Gao, S., Frejinger, E. and Ben-Akiva, M. (2010), Adaptive route choices in risky traffic networks: a Prospect Theory approach, *Transportation Research Part C: Emerging Technologies*, **18**, 727–40.

Gärling, T. and Axhausen, K.W. (2003), Introduction: habitual travel choice, *Transportation*, **30**, 1–11.

Gärling, T. and Young, W. (2001), Perspectives on travel behavior: decision paradigms. In: D. Hensher and J. King (eds), *Travel Behaviour Research: The Leading Edge*, Oxford: Pergamon.

Gigerenzer, G. and Selten, R. (2001), The adaptive tool box, in G. Gigerenzer and R. Selten (eds), *Bounded Rationality: The Adaptive Tool Box*, Cambridge, MA: MIT Press, pp. 1–12.

Helbing, D. (2004), Dynamic decision behavior and optimal guidance through information services: models and experiments, in M. Schreckenberg and R. Selten (eds), *Human Behaviour and Traffic Networks*, Berlin: Springer, pp. 47–95.

Helbing, D., Schönhof, M. and Kern, D. (2002), Volatile decision dynamics: experiments, stochastic description, intermittency control and traffic optimization, *New Journal of Physics*, **4**, 33.1–33.16.

Helbing, D., Schönhof, M., Stark, H. and Hołyst, J.A. (2005), How individuals learn to take turns: emergence of alternating cooperation in a congestion game and the prisoner's dilemma, *Advances in Complex Systems*, **8**, 87–116.

Hertwig, R., Barron, G., Weber, E.U. and Erev, I. (2004), Decisions from experience and the effect of rare events in risky choice, *Psychological Science*, **15**, 534–9.

Horowitz, J. (1984), The stability of stochastic equilibrium in a two-link transportation network, *Transportation Research Part B*, **18**(1), 13–28.

Humphrey, S.J. (2004), Feedback-conditional regret theory and testing regret-aversion in risky choice, *Journal of Economic Psychology*, **25**, 839–57.

Innocenti, A., Lattarulo, P. and Pazienza, M.G. (2013), Car stickiness: heuristics and biases in travel choice, *Transport Policy*, **25**, 158–68.

Kahneman, D. (2011), *Thinking, Fast and Slow*, New York: Macmillan.

Kahneman, D. and Tversky, A. (1979), Prospect theory – an analysis of decision under risk, *Econometrica*, **47**, 263–91.

Kahneman, D. and Tversky, A. (1984), Choices values and frames, *American Psychologist*, **39**, 341–50.

Kahneman, D., Knetsch, J. and Thaler, R. (1991), Anomalies – The endowment effect, loss aversion, and status quo bias, *Journal of Economic Perspectives*, **5**(1), 193–206.

Katsikopoulos, K.V., Duse-Anthony, Y., Fisher, D.L. and Duffy, S.A. (2002), Risk attitude reversals in drivers' route choice when range of travel time information is provided, *Human Factors*, **44**, 466.

Kemel, E. (2013), Prospect Theory for joint time and money consequences in risk and ambiguity, *Transportation Research Part B: Methodological*, **56**, 81–95.

Kim, S-S. (2011), A model of political information-processing and learning cooperation in the repeated prisoner's dilemma, *Journal of Theoretical Politics*, **24**, 46–65.

Larrick, R.P. (1993), Motivational factors in decision theories: the role of self-protection, *Psychological Bulletin*, **113**, 440–50.

Leontiadis, I., Marfia, G., Mack, D., Pau, G., Mascolo, C. and Gerla, M. (2011), On the effectiveness of an opportunistic traffic management system for vehicular networks, *Intelligent Transportation Systems, IEEE Transactions on*, **12**(4), 1537–48.

Levinson, D. (2003), The value of advanced traveler information systems for route choice, *Transportation Research Part C: Emerging Technologies*, **11**, 75–87.

Loomes, G. and Sugden, R. (1982), Regret theory: an alternative theory of rational choice under uncertainty, *The Economic Journal*, **92**, 805–24.

Lu, X., Gao, S. and Ben-Elia, E. (2011), Information impacts on travelers' route choice behavior in a congested risky network, *Transportation Research Record*, **2243**, 89–98.

Lu, X., Gao, S., Ben-Elia, E. and Pothering, R. (forthcoming), The effect of real-time information on travelers' route-choice behaviors, *Mathematical Population Studies*, forthcoming.

Luce, R.D. and Raiffa, H. (1957), *Games and Decisions*, New York: John Wiley & Sons.

Mahmassani, H.S. and Jayakrishnan, R.J. (1991), System performance and user response under real time information in a congested traffic corridor, *Transportation Research Part A*, **25**(5), 293–307.

Mahmassani, H.S. and Liu, Y.H. (1999), Dynamics of commuting decision

behaviour under advanced traveller information systems, *Transportation Research Part C: Emerging Technologies*, **7**, 91–107.
Mokhtarian, P.L. (1990), A typology of relationships between telecommunications and transportation, *Transportation Research Part A: Policy and Practice*, **24**, 231–42.
Mokhtarian, P.L. (2009), If telecommunication is such a good substitute for travel, why does congestion continue to get worse?, *Transportation Letters*, **1**, 1–17.
Nakayama, S. and Kitamura, R. (2000), Route choice model with inductive learning, *Transportation Research Record*, **1725**, 63–70.
Nakayama, S., Kitamura, R. and Fujii, S. (1999), Drivers' learning and network behavior: dynamic analysis of the driver-network system as a complex system, *Transportation Research Record*, **1676**, 30–36.
Nakayama, S., Kitamura, R. and Fujii, S. (2001), Drivers' route choice rules and network behavior: do drivers become rational and homogeneous through learning? *Transportation Research Record*, **1752**, 62–8.
Olson, M. (1965), *The Logic of Collective Action: Public Goods and the Theory of Groups*, Cambridge, MA: Harvard University Press.
Prashker, J.N. and Bekhor, S. (2004), Route-choice models used in the stochastic user equilibrium problem: a review, *Transport Reviews*, **24**, 437–63.
Rapoport, A., Kuglar, T., Dugar, S. and Gisches, E.J. (2009), Choice of routes in congested traffic networks: experimental tests of the Braess' paradox, *Games and Economic Behavior*, **65**, 538–71.
Razo, M. and Gao, S. (2011), A rank-dependent expected utility model for strategic route choice with stated preference data, *Transportation Research Part C: Emerging Technologies*, **27**,117–30.
Salomon, I. (1986), Telecommunications and travel relationships: a review, *Transportation Research Part A: Policy and Practice*, **20**, 223–38.
Schaffers, H., Komninos, N., Pallot, M., Trousse, B., Nilsson, M. and Oliveira, A. (2011), Smart cities and the future Internet: towards cooperation frameworks for open innovation, in J. Dominique et al. (eds) *Future Internet Assembly*, *Lecture Notes in Computer Science*, **6656**, 431–46.
Schwartz, S.H. (1977), Normative influences on altruism. In L. Berkowitz (ed.), *Advances in Experimental Social Psychology*, Vol. 10, San Diego: Academic Press, pp. 221–79.
Schwartz, S.H. and Howard, J.A. (1981), A normative decision-making model of altruism. In J.P. Rushton and R.M. Sorrentino (eds), *Altruism and Helping Behavior*, Hillsdale, NJ: Lawrence Erlbaum, pp. 189–211.
Selten, R., Chmura, T., Pit, T., Kube, S. and Schreckenberg, M. (2007), Commuters' route choice behavior, *Games and Economic Behavior*, **58**, 394–406.
Simon, H.A. (1982), *Models of Bounded Rationality*, Cambridge, MA: MIT Press.
Slovic, P., Finucane, M.L., Peters, E. and MacGregor, D.G. (2007), The affect heuristic, *European Journal of Operational Research*, **177**, 1333–52.
Steg, L. (2005), Car use: lust and must. Instrumental, symbolic and affective motives for car use, *Transportation Research Part A: Policy and Practice*, **39**, 147–62.
Sunitiyoso, Y., Avineri, E. and Chatterjee, K. (2011), The effect of social interactions on travel behaviour: an exploratory study using a laboratory experiment, *Transportation Research Part A: Policy and Practice*, **45**, 332–44.
Thaler, R.H. and Sunstein, C.R. (2008), *Nudge: Improving Decisions about Health, Wealth, and Happiness*, New Haven, CT: Yale University Press.

Thorndike, E.L. (1898), Animal intelligence: an experimental study of the associative processes in animals, *Psychological Monographs*, **2**(4).
Tian, L-J., Huang, H-J. and Gao, Z-Y. (2012), A cumulative perceived value-based dynamic user equilibrium model considering the travelers' risk evaluation on arrival time, *Networks and Spatial Economics*, **12**, 589–608.
Train, K. (2002), *Discrete Choice Methods with Simulation*, Cambridge, MA: Cambridge University Press.
Triandis, H.C. (1977), *Interpersonal Behavior*, Monterey, CA: Brooks/Cole.
Triandis, H.C. (1980), Values, attitudes, and interpersonal behavior. In H. Howe and M.M. Page (eds), *Nebraska Symposium on Motivation 1979*, Lincoln: University of Nebraska Press, pp. 195–259.
Tversky, A. and Kahneman, D. (1974), Judgment under uncertainty: heuristics and biases, *Science*, **185**, 1124–30.
Tversky, A. and Kahneman, D. (1992), Advances in prospect theory – cumulative representation of uncertainty, *Journal of Risk and Uncertainty*, **5**, 297–323.
Von Neumann, J. and Morgenstern, O. (1944), *Theory of Games and Economic Behavior*, Princeton, NJ: Princeton University Press.
Wardrop, J.G. (1952), Some theoretical aspects of road traffic research, *Proceedings of Institute of Civil Engineers, Part II*, **1**, 325–78.
Zeelenberg, M., Beattie, J., Van der Pligt, J. and De Vries, N.K. (1996), Consequences of regret aversion: effects of expected feedback on risky decision making, *Organizational Behavior and Human Decision Processes*, **65**, 148–58.
Zhang, Y., Stopher, P. and Halling, B. (2013), Evaluation of South-Australia's TravelSmart project: changes in community's attitudes to travel, *Transport Policy*, **26**, 15–22.

3. User requirements for the design of efficient mobile devices to navigate through public transport networks

Cristina Pronello and Cristian Camusso

1 INTRODUCTION

Public transport, supported by the development and deployment of Intelligent Transport Systems (ITS), is increasingly considered key to achieve a sustainable transport system. To speed up and coordinate the deployment of ITS in road transport and its interfaces with other transport modes, the Commission adopted an Action Plan in 2008 (EC, 2008), followed by the Directive 2010/40/EU (European Parliament and Council, 2010). In addition, the EU global positioning system, Galileo291, is expected to be operational by 2016–17; the early services started in 2014, while 18 satellites are expected in 2015–16.

Within such a framework, the research presented in this chapter resulted in the design of 'Smart-Way', an application developed in the cities of Turin (Italy) and Dresden (Germany), conceived to use Galileo to support navigation through a public transport network. At present the application is based on the NAVSTAR GPS (NAVigation Satellite Timing And Ranging Global Positioning System), but it will be switched to Galileo when this becomes available, thus bringing new developments in ICT (Information and Communication Technology) for transport, at least in Europe.[1]

The motivation for this research comes from the awareness that, without a policy to reduce automobile use, the introduction of high quality public transport does not guarantee significant modal diversion (Mackett and Edwards, 1998; Pronello and Camusso, 2011). However, a joint policy of providing high quality and reliable information could induce people to

[1] More information about EU support of Location Based Services (LBS) can be found at: http://ec.europa.eu/enterprise/policies/satnav/galileo/applications/location-based-services/index_en.htm.

consider a modal shift. The difficulty lies in persuading habitual car users that public transport could be an alternative to the car; however, with the provision of suitable traveller information the perceived inconvenience of using public transport is reduced by making it easier to plan and execute a journey (Lyons and Harman, 2002).

Smart-Way is an Advanced Traveller Information System (ATIS), an ICT for transport, which would like to bridge the gap between user needs and their behaviour, making the latter more sustainable. ATISs are data integration systems delivering accurate, reliable and timely information to travellers (Hyejung, 2009), enabling them to plan their route, estimate their travel time, and make informed decisions using real-time information (Kumar et al., 2003).

Abdel-Aty (2002) stated that it is not easy to define and quantify ATIS impacts due to the lack of real-world environments in which travellers' behaviour can be observed under the influence of ATISs. The potential of ATISs to influence mobility behaviour has hitherto rarely been researched (Gotzenbrucker and Kohl, 2011; Chorus et al., 2006). However, there have been many attempts to evaluate ATIS benefits, gathering data from various sources, predominantly from surveys but also from field observations and simulations (Williams et al., 2008). Most of the surveys concerned the effects of traffic information on car drivers, mainly commuters, to estimate user satisfaction and the effects of ATIS operation (Khattak et al., 1993; Asakura et al., 2000; HongCheng and LiJun, 2006; Chorus et al., 2006).

Instead, only a few studies explored the consequence of information on public transport (PT) ridership, notwithstanding its potential role in increasing it and improving customer satisfaction (Jou, 2001). Moreover, as observed by Pronello and Camusso (2011), there is another component in determining behaviours: daily activities and habits can result in a resistance to change, even if opinions towards modal change are favourable. In fact, the choices made regarding the daily trip represent a repeated behaviour which can gradually become a habit and this very repetition hinders the ability of people to change it (Aarts et al., 1998). Fujii and Gärling (2003) also argued that context is a true determinant of actual behaviour; however, habits remain a key aspect in modal choice (Gärling and Steg, 2007).

Real-time information is the novelty introduced by Smart-Way, one of the first smartphone applications for PT when it was developed in 2010–11. Nowadays there are a few more real-time applications, such as those developed for Zurich (ZVV, 2013), Vienna and London.

Abdel-Aty et al. (1996) reported that about 38 per cent of non-public transport users indicated that they might consider using transit if appropriate public transport information were available to them. Further, Abdel-Aty (2001), using telephone interviews in two metropolitan areas in northern California, showed that commuters seek several types of PT information to be encouraged to use transit: operating hours, frequency of service, fare, transfers, seat availability and walking time to transit stops. He also noted that the accuracy and the quality of the information to travellers are decisive (Abdel-Aty, 2002). The Transportation Research Group (2000) found that the five most important factors that users seek in PT have been ranked in the following order: reliability, travel time, convenience, cost and comfort.

People using traveller information on a daily basis are still a small portion of all travellers, but the rapid evolution of the information given to public transport users will probably increase such a proportion (Bunch et al., 2011). Initially, general and not customized information about scheduled arrival time (timetables and network maps at bus stops), was followed by the first ATISs applied to public transport, which transferred the information available at the bus stops to the operator's website. However, this information was not yet customized, and reported only estimated arrival times. In the meantime, transport companies started offering diverse public transport information services, such as emails and SMS about estimated arrival time at stops (Tang and Thakuriah, 2012). Now even more transport companies have equipped their stops with displays showing vehicles' arrivals estimated in real time.

Tang and Thakuriah (2012) compared transit ridership before and after the implementation of real-time transit information systems, concluding that their increasing use leads to PT ridership gains. However, simple before-and-after comparisons may not be enough to explain the gains since factors other than the implementation of transit information systems – such as population, fuel price, transit fare and employment levels – might influence changes in PT patronage. Thus, it would be problematic to conclude that the observed increase in ridership is a direct result of the traveller information system based on this type of study (Schweiger, 2003). If, and in what way, systems like these have an effect is highly dependent on how they are utilized by users. Obviously, this is not only a technological but also a social process which merits technology assessment (Gotzenbrucker and Kohl, 2011). Farag and Lyons (2012) showed how travel behaviour, travel attitudes and socio-demographics have the strongest effect on pre-trip PT information use for both business and leisure trips.

The research reported in this chapter aims at defining the characteristics of a real-time traveller information device by asking the users directly what features they want to really benefit from it, to use (more) PT, and to induce car drivers to divert to PT. A further aim is to define the potential business model to keep such a system updated and operational. The view of both transport companies and transport authorities completes the picture since they were asked about their requirements for strengthening users' loyalty and attracting new users, in addition to possible barriers with using Galileo-based applications on mobile devices.

The next section of this chapter describes the quali-quantitative approach employed for the data collection exercise. The results are presented in section 3, distinguished between PT users on one hand and companies and authorities on the other. Section 4 discusses the results and contrasts them with the relevant literature.

2 METHODOLOGY

The methodology comprises four steps:

- Data collection design.
- Sample selection.
- Administration of surveys.
- Data analysis.

The first step relied on a quali-quantitative approach based on two tools: the web-questionnaire and the focus group that were meant to work in parallel. Focus groups are typically used in market analysis; for details, see Krueger and Casey (2000).

The focus groups were used both to collect users' needs and investigate the effect of the technology studied on travel behaviour as well as to investigate thoroughly the psychological and social attitudes of the sample and their perception of the technology.

Different focus group outlines and questionnaires were prepared for the PT users and for the companies and authorities. Two groups of transport users were established, one in Turin and one in Dresden, to observe possible differences due to social and cultural backgrounds. European PT companies and authorities were involved in a single group.

Approximately 10 individuals per group of users were selected following a stratified convenience sampling plan according to the gender (male, female); age (<25, 26–65, >65); profession/educational level/income (low-middle, high); used mode (car, PT). Fourteen people took part in the focus group in Turin, including four physically disabled persons to gain a wider view of mobility needs. The German focus group had seven participants without any disabled people.

Ten companies and authorities were stratified by geographical location (north, south, and east of Europe) and city size (medium, large).

Many PT companies were initially contacted and asked to fill in the web-questionnaire. Given the limited availability to travel to the focus group's venue, six transport companies and one transport authority participated at the focus group:

- Transports Metropolitans de Barcelona TMB, and Athens Urban Transport Organisation OASA (large cities in southern Europe).
- Budapest Transport Private Corporation Transman, and Zarząd Transportu Miejskiego ZTM (Warsaw Public Transport Authority) (large cities in eastern Europe).
- Dresdner Verkehrsbetriebe AG DVB (from Dresden, a medium-size city in north-eastern Europe).
- Empresa Municipal de Transportes de Valencia S.A.U. EMT, and Gruppo Torinese Trasporti GTT (medium-size cities in southern Europe).

One more transport company responded to the questionnaire: Strathclyde Partnership for Transport (SPT) from Glasgow (medium-size city in north-western Europe).

2.1 Focus Group Outlines

Transport users were selected to offer input to develop the application incorporating user needs, thus the focus groups started with a presentation of the Smart-Way project and an explanation about the concept of the real-time travel planner for PT. Then, a '*tour de table*' allowed participants to meet and got the discussion going, which was divided in four parts.

The first part was dedicated to understanding the personality traits of the participants, mainly related to the emotions felt during their trips (for example ability to 'navigate' in the networks, sense of direction and attitude towards changing routes). The second part dealt with travel habits, to understand mobility behaviour and attitudes towards public transport and cars.

The third part concerned technological issues, aimed to understand the level of confidence of the participants with technologies, such as the Internet, mobile phones, computers, and contemporary tools such as social media. Participants were then asked whether they own technological devices and whether those are equipped with GPS. The last part revolved around the Smart-Way concept: the interest in using the application and the characteristics it should embody were discussed. Finally, participants were asked about which features are required of the devices to induce modal diversion or increase use of PT.

After the presentation of the project and the introduction of the participants, the transport companies and authorities focus group concentrated on understanding which information services are offered by companies, how, and who manages those. Privacy issues were also examined, because they were an important concern of users when discussing ICT for transport issues.

Before entering the 'core' of the Smart-Way application, the discussion concerned the use of GPS and its possible drawbacks, the opinions about Galileo and the willingness to use it.

Further points investigated included: the utility of using Smart-Way for the transport companies and transport authorities, their expectations, their opinion about what users expect from such a device and about its ability to attract more people to public transport, the possible privacy implications, the information they would like to obtain and their willingness to implement such a system, including the point regarding whether its management should be in-house or outsourced.

2.2 Questionnaires Supporting the Focus Groups

A web-questionnaire was administered and filled in by the transport users and PT companies and transport authorities before the focus group.

Initial questions for PT users covered personal information and were complementary to the topics of the discussion. Further questions regarded travel behaviour of users (the characteristics of their most frequent trip and of trips in their spare time) and their opinions about private and public transport, and about technological tools, expressed according to a five-point Likert scale (Table 3.1).

The web-questionnaire for PT companies and transport authorities started with requests for general figures (number of employees, revenues, territorial coverage, used modes and fleet), followed by questions about information services: the characteristics of their system in terms of the time release of the data (static or dynamic), the time interval in the case of real-time information, the kinds of services offered (SMS, WEB, info point, call centre) and the costs of offering them, as well the data given to users. Finally, questions about privacy issues explored potential drawbacks of the Smart-Way implementation, highlighting again any concerns about this matter, which may be related to the concerns discussed in Chapter 9.

Table 3.1 Extract from the questionnaire administered to transport users: attitudinal/perceptive section and use of technological tools section

ATTITUDINAL AND PERCEPTIVE

Please give your opinion on the following statements

11) I like driving cars

1 Not at all	2	3	4	5 Very much

12) I like commuting by car

1 Not at all	2	3	4	5 Very much

13) I like using Public Transport

1 Not at all	2	3	4	5 Very much

14) If public transport was free of charge, I would avoid using the car

1 Not at all	2	3	4	5 Very much

15) If public transport was free of charge, I would still use the car, but I would use Public Transport more

1 Not at all	2	3	4	5 Very much

16) If public transport was free of charge, I would not change my habits

1 Not at all	2	3	4	5 Very much

17) Public transport suits my transport needs well

1 Not at all	2	3	4	5 Very much

Table 3.1 (continued)

USE OF TECHNOLOGICAL TOOLS
Please give your opinion on the following statements

25) Do you feel confident with technology?

1 Not at all	2	3	4	5 Very much

26) Do you enjoy using new technological tools/instruments?

1 Not at all	2	3	4	5 Very much

27) Are you usually updated on new technologies?

1 Not at all	2	3	4	5 Very much

28) Do you usually change your devices for the newest ones?

1 Not at all	2	3	4	5 Very much

29) Please specify the relevance of cost when you choose to buy an electronic device

1 Not at all	2	3	4	5 Very much

30) Please specify the relevance of design when you choose to buy an electronic device

1 Not at all	2	3	4	5 Very much

31) Please specify the relevance of fashion when you choose to buy an electronic device

1 Not at all	2	3	4	5 Very much

32) Please specify the relevance of user-friendliness when you choose to buy an electronic device

1 Not at all	2	3	4	5 Very much

Table 3.1 (continued)

33) Please specify the relevance of reliability when you choose to buy an electronic device

1 Not at all	2	3	4	5 Very much

34) Please specify the relevance of durability when you choose to buy an electronic device

1 Not at all	2	3	4	5 Very much

35) Please specify the relevance of being most up-to-date when you choose to buy an electronic device

1 Not at all	2	3	4	5 Very much

36) Please specify the relevance of other eventual characteristics, if applicable, when you choose to buy an electronic device: please specify which characteristic: . . . .

1 Not at all	2	3	4	5 Very much

2.3 Data Analysis Method

The discussions during the focus groups were recorded by both audio and video and subsequently transcribed. The transcriptions were then carefully read in order to draw a synoptic grid including main themes and sub-themes, thus creating the structure for the content analysis. Then the participants' wordings on the different topics were reported in the grid. This work was carried out iteratively, to organize raw data in a definite structure (Krueger and Casey, 2000). The data collected through the questionnaires were analysed together for the Italian and German groups, to better show similarities and differences and check whether the geographical location, in conjunction with different cultural habits, makes people behave differently. Only descriptive statistics were used due to the small size of the sample.

3 RESULTS

3.1 Transport Users

The 21 participants of the Italian and German focus groups were not selected to represent the Turin or Dresden population, but to include different users' profiles so as to better test all possible reactions. However, comparing the characteristics of the two samples with those of the population of their two cities (ISTAT, 2012; Demographie konkret on line, 2013), it can be observed that the Italian sample is biased only in terms of educational level (65 per cent of participants hold a university degree whereas only 12 per cent of the residents do) (ISTAT, 2013), while the German sample is biased only in terms of gender balance (42 per cent were women compared to the national average of 51 per cent) and age (no elderly people, compared to 26.8 per cent in Dresden).

Most of the participants are single, and the most frequently occurring household size is 2 people (2.2 in Dresden and 2.13 in Turin). Table 3.2 offers an overview of 19 out of the 21 participants' socio-economic characteristics and travel habits; two disabled persons are not included as they did not respond to the questionnaire.

In general, the German participants use public transport more than the Italian ones, preferring tram and bus services for both their most frequent (almost daily) and leisure journeys. The Italian participants are tied to their car and unwilling to divert even in the event of free-of-charge public transport. This is confirmed by national and regional statistics; in fact, Italians do not use public transport often, and usually do not have a yearly pass, but prefer monthly or weekly options as shown by the observatory (Audimob) of the Italians' travel behaviour (Isfort, 2011), which reports a modal share of 11.4 per cent for PT and 65.6 per cent for private cars. Comparing the two cities, the EPOMM data (2013) show a modal split for the car of 64 per cent in Turin versus 38 per cent in Dresden. PT modal shares are closer: 28 per cent in Turin versus 21 per cent in Dresden. The details about the PT services are given later in Table 3.4.

Table 3.3 presents the answers to the questions on travel habits, opinion on PT, attitude towards technology and GPS, and willingness to pay (WTP) to save time. WTP was examined asking the question 'Suppose you could complete your MOST FREQUENT TRIP saving 20 per cent of your actual travel-time. Which monetary value would you assign to the saved time?' (choice among: None; 0.5 €; 0.6 – 1.5 €; 1.6 – 3 €; 3.1 – 5 €; More than 5 €).

A final open question gave respondents the opportunity to indicate which features the Smart-Way application or device should have in order to attract car users to public transport.

The participants think that in order to attract users to PT, Smart-Way should be an easy-to-use application working on mobile phones, giving fast, reliable and real-time information about all possible trip solutions, public transport network connections, tickets' costs and parking. Users want user-friendliness, and German respondents ask for a large display to adequately show the network; the timetable; the routes; the stops; the position along the route of the bus they are travelling on, or of the vehicle which is marked as the next departure; information on connections at the stop where they get off a vehicle or the distance to the stop where they will find a connecting service. Finally, they ask for information about tickets and the option to buy them, because 'a lot of car-drivers are not able to use a ticket machine, and looking for a convenient ticket can be annoying'.

The Italian respondents require a multilingual application that is interactive with central assistance.

Both the Italian and German participants need the application to locate them, give their route plan with departure and arrival times, and they stress the importance of having the best PT solution in real time, taking into account waiting times and accidents. That way they expect to save a lot of time by PT in rush hours and declare that public transport could be competitive in terms of cost and travel time because of reserved bus lanes and because public transport can enter the area where car access is restricted (in central Turin).

German respondents require features assisting with travelling by modes other than PT: flexible guidance to their target with different means of transport; the ability to guide car drivers who are unfamiliar with local public transport. They want an application that is: 'easy to use like a navigation system to follow their entrenched habit'.

Only two Italian participants believe that improving PT quality (increasing reliability and frequency) could contribute to modal diversion more than ITS services. They also ask for a low priced application and for the option to purchase it for a limited number of days.

The discussion confirmed the preference for car use by the Italian group; they feel more stressed than Germans while travelling, especially when they drive for work; however, they think the car is their only choice due to its greater flexibility. This is the main reason they drive instead of using public transport. The latter is seen as unreliable in terms of arrival time and not acceptable for work-related travel.

Both German and Italian participants using PT declared that they do not feel stressed while travelling. However, they noted problems related

Table 3.2 Main socioeconomic characteristics and trip habits of the users (not including disabled people)

Transport mode used in the most frequent trip		On foot	Bus Tram Car driver	Car driver	Metro Train Taxi	Bus/ Tram	Metro, Train Car (driver passenger)
Number of modes owned by household	Bike	4	1	0	0	2	0
	Motorbike	1	0	0	0	0	0
	Car	3	2	2	1	1	2
Family size		4	3	2	1	3	2
Monthly income level	Household[1]	2	3	3	3	3	3
	Individual[2]	n.d.	n.d.	3	5	2	2
Occupation		Student	Student	Profes-sional accoun-tant	Free-lance advisor	Uni-versity employee	Responsible of project commu-nication environment
Education[3]		Degree	Degree	Degree	Degree	PhD	Degree
Age		26	26	41	52	47	27
Sex		M	F	F	F	F	F
Users code		User 5	User 6	User 7	User 8	User 9	Users 10
Country		Italy	Italy	Italy	Italy	Italy	Italy

Notes:

1. Household monthly income level: 1 = 'Up to 1000€'; 2 = '1–2000€'; 3 = '2–5000€'; 4 = '5–8000€'; 5 = '8–12000€'; 6 = 'more than 12000'.
2. Personal monthly income level: 1 = 'Up to 1000€'; 2 = '1–1500€'; 3 = '1.5–2000€'; 4 = '2–2500€'; 5 = '2.5–3000€'; 6 = '3–4000€'; 6 = 'more than 4000€'.
3. Education: SS = 'Secondary School'; HSD = 'High School Diploma'.

to connections, ticket machines (especially for German participants) and crowded vehicles during rush hours.

The majority declared that when they plan a trip they start a day before the departure using Internet services and public transport websites. When travelling in unknown cities they use maps and information points. All the users stated they are not afraid of getting lost in an unknown city because they believe they have a good sense of direction or they can easily obtain help to find the right way.

Usability findings

Whilst discussing Smart-Way, everybody declared that they prefer a mobile application rather than a new device. Such an application

Table 3.2 (continued)

Bus Tram Train	n.d.	n.d.	Car driver Bicycle	Bus Tram On foot	Bus Tram Metro Train Taxi Car driver	Bus/ Tram Train, Car driver	Bus Tram On foot
2	2	1	2	2	2	2	3
3	0	0	0	0	0	0	0
3	3	1	2	1	2	1	0
4	2	2	3	2	1	3	2
3	3	4	3	2	n.d.	3	3
3	n.d.	7	n.d.	3	n.d.	2	6
Free-lance advisor	Retired	Retired	Student	Employee	n.d.	Employee adminis-tration	Consult-ant/ engineer
Degree	Degree	Degree	Degree	Degree	HSD	HSD	Degree
32	64	59	27	58	40	24	31
M	F	M	M	M	M	M	M
User 11	User 12	User 13	User 14	User4208	User4209	User4210	User4211
Italy	Italy	Italy	Italy	Germany	Germany	Germany	Germany

could be useful for leisure trips and in unknown cities, and those are probably the only cases when they are willing to spend money for the service.

The main suggestions about Smart-Way features are:

- Using the device should be low cost, also considering that fees for Internet connections are different around Europe.
- Users prefer to buy and download the application only for the time needed, thus paying for it according to an hourly rate or for a specific period: day, week, month. Additionally, it should be possible to buy and download the application using Bluetooth or infrared, for example in airports, main railway and bus stations, information points.
- A market segmentation is preferable, distinguishing the business or everyday users from the tourists. Regarding leisure time, the Smart-Way application could be distributed as a CD with the software needed for the cell-phone or together with a traditional tour guide.

Table 3.2 (continued)

Transport mode used in the most frequent trip		Bus Tram Car driver On foot	Bus Tram	Bus Tram	Bus Tram Car (driver/ passenger)	Bus Tram Train, Car passenger, On foot
Number of modes owned by household	Bike	1	2	5	3	4
	Motorbike	0	1	0	0	0
	Car	1	0	2	1	1
Family size		2	2	6	2	4
Monthly income level	Household[1]	3	2	3	3	3
	Individual[2]	6	1	2	4	1
Occupation		German professional providing auditing services	Appren-ticeship/ trainee industrial clerk	Paralegal (working at a lawyer`s)	Research assistant	Voluntary year of social service
Education[3]		Degree	SS	HSD	PhD	SS
Age		42	19	33	57	17
Sex		M	F	F	M	M
Users code		User4212	User4213	User4214	User4215	User4216
Country		Germany	Germany	Germany	Germany	Germany

- Smart-Way must be useful for optimizing travel time.
- It could be useful to locate children, sending a SMS to their device, which should then answer automatically, reporting its position.
- The system should be scalable and compatible with all cell-phones, interacting with all the interesting information and adding new data (restaurants, hotels, other services, etc.) when they become available.
- Users would like to buy tickets simply using the mobile phone.

The desired characteristics of the Smart-Way device were as follows:

- Easy to use.
- Large screen, large buttons.
- Option to change languages, large fonts for the elderly.
- Low-battery consumption.

Considering the possibility of a modal diversion caused by Smart-Way, all the participants envisaged a potential, although they expressed concerns

Table 3.3 Focus group transport users: travel habits, PT opinions and attitudes towards technology as declared in the quantitative questionnaire

Characteristics	Italian users	German users
Most frequent trip: reasons for car use[1]	1. Inadequate coverage of PT network, flexibility, other 2. Time constraints, habits, comfort	1. Time constraints, comfort, other 2. Inadequate coverage of PT network, route constraints, habit, personal safety
Most frequent trip: reasons for PT use[1]	1. Time constraints, cost 2. Good coverage of PT network, route constraints, other 3. Good PT service (frequency), route safety, personal safety	1. Cost, good coverage of PT network, good PT service (frequency) 2. Time constraints, other 3. Habit
Transport mode used in leisure trips[1]	1. Car (as driver), car (as passenger) 2. On foot 3. Bus/tram 4. Metro, train	1. Bus/tram 2. On foot 3. Car (as driver), car (as passenger) 4. Train, taxi, bicycle 5. Metro, other
Declared reasons for using car in general[1]	1. Time constraints 2. Inadequate public transport network coverage 3. Route constraints, flexibility, comfort 4. Inadequate PT service (frequency) 5. Habit, personal safety, other	1. Time constraints, habit 2. Inadequate Public Transport service (frequency) 3. Comfort, route constraints, inadequate public transport network coverage 4. Personal safety, other 5. Habit, costs
Reasons for using PT in general[1]	1. Car parking unavailability 2. Costs, time saving, other 3. Personal safety, health/environment, comfort	1. Costs, car parking unavailability, health/ environment 2. Habit 3. No car availability, time saving, road safety 4. Comfort 5. No driving licence, personal safety

Table 3.3 (continued)

Characteristics	Italian users	German users
Weakness of PT services in your city [1]	1. Reliable time schedule 2. Time flexibility 3. Comfort, travel time 4. Cleanliness, inadequate user information, difficult park and ride 5. Disadvantages in moving heavy and large items	1. Disadvantages in moving heavy and big goods, inadequate PT service (frequency), costs 2. Time flexibility 3. Inadequate coverage of PT network, reliable time schedule, travel time 4. Difficult park and ride, inadequate user information, cleanliness, comfort
Distance and duration of the most frequent trip (declared by the users)	User5: 0.8 km; 10 minutes User6: 4 km; 12 minutes User7: 20 km; 60 minutes User8: 5 km; 20 minutes User9: 36 km; 30 minutes User10: 15 km; 20 minutes User11: 65 km; 150 minutes User12: 4 km; 20 minutes User13: 4 km; 30 minutes User14: 12 km; 20 minutes	User4208: 4 km; 30 minutes User4209: 165 km; 180 minutes User4210: 12 km; 30 minutes User4211: 1 km; 7 minutes User4212: n.a.; 75 minutes User4213: 7 km; 20 minutes User4214: 6 km; 30 minutes User4215: 16 km; depends User4216: 20 km; 8 minutes
User approach and confidence towards technology	Less confident with technology. They like to upgrade their device but do not like to change them frequently. Most users own a device such as a mobile phone, digital camera and use a service such as the Internet and social media. 50% of users have a GPS navigator	Confident with technology, they do not enjoy using this kind of device. They like to upgrade their devices but do not like to change them frequently. Most of the users own a device such as a mobile phone, digital camera, Internet and social network. 33% of users have a GPS navigator
Most frequent trip: WTP for 20% time reduction	1. None (5 users) 2. 0.50€ (1 user), 0.6–1.5€ (1 user); 1.6–3€ (1 user); 3.1–5€ (1 user); more than 5€ (1 user)	1. None (6 users) 2. 0.50€ (1 user), 0.6–1.5€ (1 user); 3.1–5€ (1 user)

Note: 1. The answers are reported in order of importance from the most chosen to the least.

and doubts. In fact, a device helping users to save time – especially professionals with time limitations – is not enough to bring about a real change if public transport services are not improved in terms of frequency, trip time and reliability. Moreover, saving time is not a main requirement for leisure trips, and people on holiday could be unwilling to pay for Smart-Way.

Thus, market segmentation, both in terms of transport users' profile and in terms of travel purpose, is mandatory to achieve market penetration.

Willingness to pay (WTP) is also different for users of different types and from different countries:

- German participants would spend money once only for the device or the application (about 30 euros), and would like to use it without any further payment. Regarding a fee, they think it would be reasonable to spend 20–30 per cent of the overall cost of the tickets. Some Italian respondents stated that they were willing to spend from 5 to 10 euros per month for the application while others declared a willingness to spend from 7 to 8 per cent of their actual monthly cell-phone bills.
- Italian pensioners in the sample, not having any time constraints, are unwilling to spend money. They think that Smart-Way should be given free of charge by municipalities or PT companies, because it is a tool that could encourage people to use public transport more. In other cases, the suggestion is to include Smart-Way fees in the PT pass price.

Regarding privacy issues, a lot of participants were aware of the possibility to be localized with GPS devices and therefore also with the Smart-Way application. While German participants do not have privacy concerns, some Italian respondents wish to be able to switch off the device in order to decide whether to be located. These privacy and surveillance-related issues are further discussed in Chapter 9.

Focusing on particular groups

The point of view of physically disabled people is very different. They have high expectations of the Smart-Way project and gave precise suggestions. When asked about their feeling towards travelling, their perception was clear: fear, dread, anxiety. Physically disabled people like moving in their city and also abroad, but are worried since they are not sure whether public transport will take them home. Being able to move is a great freedom for them and affects their quality of life.

Physically disabled participants are confident with technology because it helps their everyday life: PCs and the Internet are used on a daily basis for both work and leisure. When they want to make a trip, they have to plan it in advance:

- In Turin, not all PT vehicles are equipped for passengers who use wheelchairs, and they often have to book a special service two days in advance.
- Concerning leisure trips in Italy or abroad, travellers with disabilities have to book in advance special train services and assistance for train or air travel and even the travel destination is chosen taking into account their specific needs.

All those tasks are time consuming because of bureaucracy, and Smart-Way could reduce this problem, offering all the required information in real time.

For those reasons, people with disabilities require information such as:

- Whether the vehicle arriving is equipped with the facilities to accommodate them or when such a vehicle will arrive.
- The characteristics of bus stops and their equipment: if there is a ramp, if the bus stop platform is at the same level as the bus floor.
- The functioning of elevators in metro stations.
- The presence of special parking near bus, metro and train stations.

All such information is essential for physically disabled people and they need it before their departure and during the whole duration of the trip. The device should be extremely easy to use, with a few buttons only. Touch-screens are not so useful: there are people with hand mobility problems who struggle with the precision required for touch-screen tools. In addition, they also prefer that the device is a mobile application rather than an additional device.

If Smart-Way were to help physically disabled people with their needs, they would be willing to spend money for it: 100–200 euros for the application with all the information or a monthly fee from 5 to 30 euros.

3.2 Transport Companies and Authorities

Table 3.4 reports some data about territorial coverage and profile of the eight PT companies and transport authorities who answered the questionnaire. An overview reveals that they differ regarding their size (number and length of lines, employees), territorial coverage, available budget and service types. Some of those managing different transport modes and services on an urban and metropolitan scale use a single control centre, whereas others operate several centres.

Table 3.5 presents the main outcomes of the questionnaire, showing a variety of technological equipment to acquire and distribute data as well as the different types and quality of real-time information. Currently available technological equipment is key when developing ICT for transport applications, so this comparative table is useful when designing such policies at a European-wide level.

The focus group confirmed the differences among the companies: some of them collect all the data about their services in an automated way, whereas others still do it manually, so the Smart-Way application should consider that in some cities insufficient data management platforms may exist.

The main characteristics that PT companies expect from Smart-Way are the following:

- It should be a mobile application compatible with all the mobile operators and with the standards of all European cities; it could be a multi-platform application showing standard information together with a real-time service giving more information than Google Maps.
- It should assist companies with recording data concerning their fleet.
- It should be conceived as a pedestrian travel aid, giving directions to reach stops on foot, including information on inclines and providing information about alternative routes.

Regarding potential barriers to using Smart-Way, the PT companies are able to provide all the useful information to the Smart-Way application, but want to filter it before it is distributed to users. They would prefer managing the Smart-Way system in-house to be sure of disseminating only appropriate and useful information. Moreover, many PT companies use GPS for their service management: thus they do not see problems with using the system, and the precision of the signal and the data acquisition are adequate for their needs. The main problem related to GPS is the different equipment on different vehicles of the fleet, which varies according to the vehicles' age. A weakness of some on-board GPS systems is that the position of the vehicles is relayed to the control centre by radio. This is an interesting issue related to development of ICT for public transport.

Finally, PT companies do not have any particular opinion of the Galileo system, and express some scepticism about when this system will become operational. The main use of the new Galileo system would be to replace the traditional system, using the odometer as a back-up when the present GPS does not work.

Table 3.4 PT organization profile

City	Organization	PT service territorial extension					
		City area		Metropolitan area		Regional area (if the PT service also covers this area)	
		Surface (km^2)	Inhabitants (mn)	Surface (km^2)	Inhabitants (mn)	Surface (km^2)	Inhabitants (mn)
Barcelona	TMB (private company)	98	1.6	319	2.8		
Budapest	Transman (private company)	525	1.7	Budapest Transport Association covers 192 settlements + Budapest but in some papers the agglomeration is much smaller		1215	2
Valencia	EMT (public company)		0.8		1.8		
Athens	OASA (public company)	50	0.65	544	4	1450	
Dresden	DVB (public company)	328	0.5	n.a.	n.a.	100	69 700
Torino	GTT (public company)	130.3	0.9	1127	1.7	25 399	4.4
Warsaw, Mazovia, Poland	ZTM (public company)	517.2	1.7	2279	2.4		
Glasgow / West of Scotland	SPT (public company)	177	0.58	3397	1.8	9310	2.2

Table 3.4 (continued)

Organization role	Number of employees	Information about organization's financial conditions		
		Fares revenues (k€/year)	Public subsidy (k€/year)	TOTAL (k€/year)
Metro and main urban bus operator	Bus company: 4197; Metro company: 3703	340 000	306 000	646
Operator	11839 employees (Full-time headcount figure in 2008). Other personnel engaged 728	252 000	191 000	442
Metro and main urban bus operator	1652	44 000	64 000	108
Operator	Authority (207), Operators (11 858)	272 000	140 000	411
Operator	1673	101	DVB AG is a part of Dresden's public holding. Profit and shortage will be set off in this holding.	n.a.
Planning, organization, coordination and control of the P.T. system	5500	n.a.	n.a.	n.a.
Public transport provider in Dresden	533	142	–	141
Missing	700	52 000	117	168

Table 3.4 (continued)

City	Mode	Number of vehicles	Number of lines	Length of lines [km]	Tot km of the service [km per year]	Passengers per year [millions]	Passenger* km per year [millions]
Barcelona	BUS	1080	108	923.92	42 221 000	361.7	14 390.30
	METRO	791	8	96.7	79 044 000	189.9	3520.70
Budapest	BUS	1409	213	819	1 676	546	2662
	TROLLEY BUS	167	16	66	73	77	222
	TRAM	607	32	154	231	333	943
	METRO	392	3	35	31	297	1220
	SUBURBAN RAILWAY	294	4	103	239	55	499
Valencia	BUS	480	63	939	20 400 000	100	5
Athens	BUS	2145	324	3646	113 063 000	419	n.a.
	TROLLEY BUS	366	23	195	12 019 000	92.2	n.a.
	TRAM	35	3	50	2 341 000	19.6	n.a.
	METRO	95	3	148	40 996 000	320.3	n.a.
	SUBURBAN TRAIN	n.a.	2	n.a.	n.a.	3.5	n.a.
Dresden	BUS	154	28	311.4	13.3	n.a.	199.2
	TRAM	196	12	204.5	13.5	n.a.	493
	MOUNTAIN	4	2	0.547	n.a.	n.a.	n.a.
	RAILWAY	5	4	0.274			
	FERRIES						
Torino	BUS	1180	80	1109	56	n.a.	n.a.
	TRAM	220	8	74.3	n.a.	n.a.	n.a.
	METRO	33	1	13	n.a.	144 (2013)	n.a.
Warsaw, Mazovia, Poland	BUS	1479	243	3952	114 084 910	396.6	3.48
	TRAM	780	26	360	51 368 422	223.2	4.35
	METRO	162	1	23.1	25 404 070	220.8	8.69
	RAPID URBAN RAIL SKM	18	1	37.4	4 033 805	30.9	7.65
Glasgow / West of Scotland	BUS	50 (+leased vehicles)	133	90 500	4 063 000	3.055	0.752
	METRO	41	2	10	n.a.	14.100	n.a.

4 DISCUSSION

The literature presented in section 1 shows that travellers respond to travel information in diverse ways and that it is difficult to quantify the impacts of ATISs on travellers' behaviour. The quali-quantitative approach (questionnaire and focus group) followed in this research proved very fruitful. In fact – notwithstanding the small sample – the research obtained the user needs for the Smart-Way application as well as several key aspects about the possible market segmentation and business model.

To satisfy the needs expressed by the users, Smart-Way offers innovative features to satisfy technological challenges, as shown in the screenshots in Figure 3.1, namely:

- Matching of several data sources (GIS, timetable information, real-time information on vehicles and disturbances) and corresponding design of interfaces.
- 'Non via'-navigation in case of delays or disturbances to identify alternative connections by incident-dependent weighting of nodes or edges of the PT network.
- Location of vehicles, re-routing of vehicles and inaccurate tracking results are recognized and compensated in the GIS line network in real time.
- Passenger status re-estimation in case of deviation from the route, by matching the user to the GIS or to a vehicle based on the current GPS position; or in case of lost user position (e.g. no GPS available in metro tunnel) by using the smartphones' inertial sensor data to decide whether the user is in a vehicle, walking or waiting at a stop.

The findings confirmed the users' requirements for PT traveller information systems obtained in previous studies (Harris and Konheim, 1995; Abdel-Aty, 2002) and the concerns in Schweiger (2003), Gotzenbrucker and Kohl (2011) and Tang and Thakuriah (2012). A PT real-time information service will never make a real difference without major improvements in the quality of PT services. This point emerges clearly from the different travel behaviour of German and Italian users: PT is less used in Italy also because its quality is lower than in Germany, as became clear from the questionnaires and the focus group discussion. Also the cultural attitude towards PT is different in the two countries: the Italians are particularly tied to their cars and think that nothing can be better.

Compared to previous studies, the work discussed here obtained additional information for designing a customized device and make it a tool to encourage a diversion from car to PT. For example, users prefer a mobile

Table 3.5 Technological equipments and information given by PT companies

City (Organizations)	BARCELONA (TMB)		BUDAPEST (Transman)	VALENCIA (EMT)	ATHENS (OASA)
Data recorded about the fleet position	GPS Position and time in order to calculate deviation between real and planned situation		The data are collected for operational traffic management; the data are stored in the central database	All information	Static view of daily programme
System used for data recording	Operational Assistance System (OAS) for buses; Regulation System (RS) in the metro		Automatic Vehicle Monitoring System of OTE (IT) on 44% of the bus fleet.	GPS / HSDPA 36+	Stationmaster uses mobile phone to obtain and record fleet position
Fleet position real time data	Continuous (real time)	Intermittent (discrete)	Intermittent (discrete)	Continuous (real time)	n.a.
Delay or time interval between data transmission	About 15 sec	30 sec	The vehicles are tracked when they reach a checking point. The 'real time' data are archived on daily basis	15 sec	n.a.
Recorded information about service characteristics (time of arrival, on-time arrival, delays, trip duration, etc.)	Arrival time, on-time arrival, delays, trip duration, fuel consumption, sales on-board, ticket validation, vehicle diagnosis		Time of arrival, on-time arrival, delays	Schedule, frequency, position, ticketing, accidents	Arrival time, departure time, delays, trip duration, percentage of completed routes

Table 3.5 (continued)

City (Organizations)	DRESDEN (DVB)		TURIN (GTT)	WARSAW (ZTM)	GLASGOW (SPT)
Data recorded about the fleet position	Position, time stamp, car ID number, car model/type, car features (wheelchair ramp, information system for unsighted), destination, route number, tour/circulation number		n.a.	Transport Authority cannot locate individual vehicles. Each carrier has its own communication and positioning system	Time, location (nearest to postcode area, vehicle registration number, current speed, mph), direction, map location
System used for data recording	Computer aided operations control system of producers CSC and TRAPEZE		n.a.	ZTM does not use any system to obtain and record fleet position information	Two separate GPS tracking systems (including mobile data terminals)
Fleet position real time data	Continuous (real time)	Intermittent (discrete)	Intermittent (discrete)	n.a.	Continuous way (real time)
Delay or time interval between data transmission	Less 1 sec.	Once in 15 sec, around the stops less than 15 sec (dynamic polling)	n.a.	n.a.	3 minutes for DRT services, 5 minutes for subsidized bus services
Recorded information about service characteristics (time of arrival, on-time arrival, delays, trip duration, etc.)	The system permanently compares schedule and performance to get: in schedule, delay, early arrival		n.a.	ZTM gets information about arrival time, on-time arrival, delays, trip duration	Arrival time, on-time arrival, time delays, trip duration

Table 3.5 (continued)

City (Organizations)		BARCELONA (TMB)	BUDAPEST (Transman)	VALENCIA (EMT)	ATHENS (OASA)
Systems used to get and record time characteristics of the service		SAP	Automatic Vehicle Monitoring System of OTE (IT) on 44% of bus fleet	Business Intelligence (Oracle)	Manual counts by crew using forms
Time data typologies		Intermittent (discrete)	n.a	Continuous (real time)	n.a.
Delay or time interval between data		Depends on the data needed; between 30 sec and 24h.	n.a.	15 sec	n.a.
INFORMATION SERVICES	SMS service (user can ask for information about arrival time at stops)	Yes for some bus lines	–	Yes	–
	WEB service for online travel planning	Yes	–	Yes	Yes
	WEB service for getting real time information	Yes	–	Yes	–
	WEB service for online ticket purchasing	–	–	Yes	–
	Info point	Yes	Yes independent from TMC	Yes	–
	Call centre	Yes managed externally by the City Hall of Barcelona	Yes	Yes	Yes
	Other (specify)	–	–	–	–

Table 3.5 (continued)

City (Organizations)		DRESDEN (DVB)		TURIN (GTT)	WARSAW (ZTM)	GLASGOW (SPT)
Systems used to get and record time characteristics of the service		Computer aided operations control system of producers CSC and TRAPEZE		n.a.	Information is available for employees on the server	Same system as above (GPS tracking)
Time data typologies		Intermittent (discrete)	Continuous (real time)	n.a.	Intermittent (discrete)	Continuous (real time)
Delay or time interval between data		Once in 15 sec, around the stops less than 15 sec (dynamic polling)	Less 1 sec	n.a.	Data are loaded with variable time interval	3 min for DRT services, 5 min for subsidized bus services
INFORMATION SERVICES	SMS service (user can ask for information about arrival time at stops)	Yes		Yes	–	Yes
	WEB service for online travel planning	Yes		Yes	Yes	Yes
	WEB service for getting real time information	Yes		–	Yes	Yes
	WEB service for online ticket purchasing	–		Yes	–	Yes
	Info point	Yes		Yes	Yes	Yes
	Call centre	Yes		Yes	Yes	Yes
	Other (specify)	Ticket purchasing for mobile phones		–	–	–

Table 3.5 (continued)

	City (Organizations)	BARCELONA (TMB)	BUDAPEST (Transman)	VALENCIA (EMT)	ATHENS (OASA)
TELEMATIC INFORMATION GIVEN TO USERS AT STOPS/STATIONS	Timetabled arrival/ departure times at stops	Yes	–	Yes	–
	Real time arrival/ departure times at stops	Yes for metro and partially for bus	Yes but few stops	Yes	–
	Service frequencies	Yes	–	–	Yes
	Unexpected service break (e.g. caused by an accident)	Yes for metro and partially for bus	–	Yes	Yes
	Planned service variation (e.g. caused by road works)	Yes	–	Yes	Yes
	Other (specify)	–	–	Marketing	–
ON-BOARD INFORMATION GIVEN TO USERS Visual information	Which is the next stop	Yes for metro and partially for bus	Yes	Yes	Yes
	Time arrival for the next stop	–	–	–	–
	Connection with other lines	Yes for metro and partially for bus	Yes	Yes	Yes
	Unexpected service break (e.g. caused by an accident)	Yes for metro and partially for bus	–	–	–
	Planned service variation (e.g. caused by road works)	Yes	–	Yes	–

Table 3.5 (continued)

	City (Organizations)	DRESDEN (DVB)	TURIN (GTT)	WARSAW (ZTM)	GLASGOW (SPT)
TELEMATIC INFORMATION GIVEN TO USERS AT STOPS/STATIONS	Timetabled arrival/ departure times at stops	Yes	Yes	–	Yes
	Real time arrival/ departure times at stops	Yes	Yes	Only tramway in Al. Jerozolimskie and underground railway	Yes
	Service frequencies	–	Yes	–	Yes
	Unexpected service break (e.g. caused by an accident)	Yes	–	–	Yes
	Planned service variation (e.g. caused by road works)	Yes	–	–	Yes
	Other (specify)	–	–	–	–
ON-BOARD INFORMATION GIVEN TO USERS Visual information	Which is the next stop	Yes	Yes	Yes	–
	Time arrival for the next stop	–	–	–	–
	Connection with other lines	Yes	–	–	–
	Unexpected service break (e.g. caused by an accident)	Yes	–	–	–
	Planned service variation (e.g. caused by road works)	Yes	–	Yes	–

Table 3.5 (continued)

	City (Organizations)	BARCELONA (TMB)	BUDAPEST (Transman)	VALENCIA (EMT)	ATHENS (OASA)
ON-BOARD INFORMATION GIVEN TO USERS Audio information	Which is the next stop	Yes for metro and partially for bus	Partly	Yes	Yes
	Time arrival for the next stop	–	–	–	–
	Connection with other lines	Yes for metro and partially for bus	Yes	–	Yes
	Unexpected service break (e.g. caused by an accident)	Yes for metro and partially for bus	–	–	–
	Planned service variation (e.g. caused by road works)	–	–	Yes	–
ANNUAL COST	SMS services (€/year)	40 000 Cost shared between SMS & Telematic info at stops	–	15 000 sms, price shared between customer and EMT (50%)	n.a.
	WEB services (€/year)	130 000	–	40 000	1000
	Info point (€/year)	0	–	60 000	87 000
	Call centre (€/year)	56 000	–	200 000	587 407
	Telematic info given to users at stops (€/year)	Cost shared between SMS & Telematic info at stops	–	55 000	n.a.
	On-board information given to users (€/year)	0	–	Free: BUSSI system is a commercial channel paid to EMT (it also installs hardware and TFT screens onboard)	n.a.
	Other (€/year)	226 000	–	–	32 120: Maps

Table 3.5 (continued)

	City (Organizations)	DRESDEN (DVB)	TURIN (GTT)	WARSAW (ZTM)	GLASGOW (SPT)
ON-BOARD INFORMATION GIVEN TO USERS Audio information	Which is the next stop	Yes	Yes	Yes	Yes
	Time arrival for the next stop	–	–	–	Yes
	Connection with other lines	No line numbers, only means of transport	–	–	–
	Unexpected service break (e.g. caused by an accident)	By driver	–	–	–
	Planned service variation (e.g. caused by road works)	By driver	–	–	–
ANNUAL COST	SMS services (€/year)	>145 000	n.a.	–	–
	WEB services (€/year)	6000	n.a.	1200	5000
	Info point (€/year)	n.a.	n.a.		70 000
	Call centre (€/year)	n.a.	n.a.	11 200	n.a.
	Telematic info given to users at stops (€/year)	n.a.	n.a.		
	On-board information given to users (€/year)	n.a.	n.a.	–	n.a.
	Other (€/year)	n.a.	n.a.	930: website	–

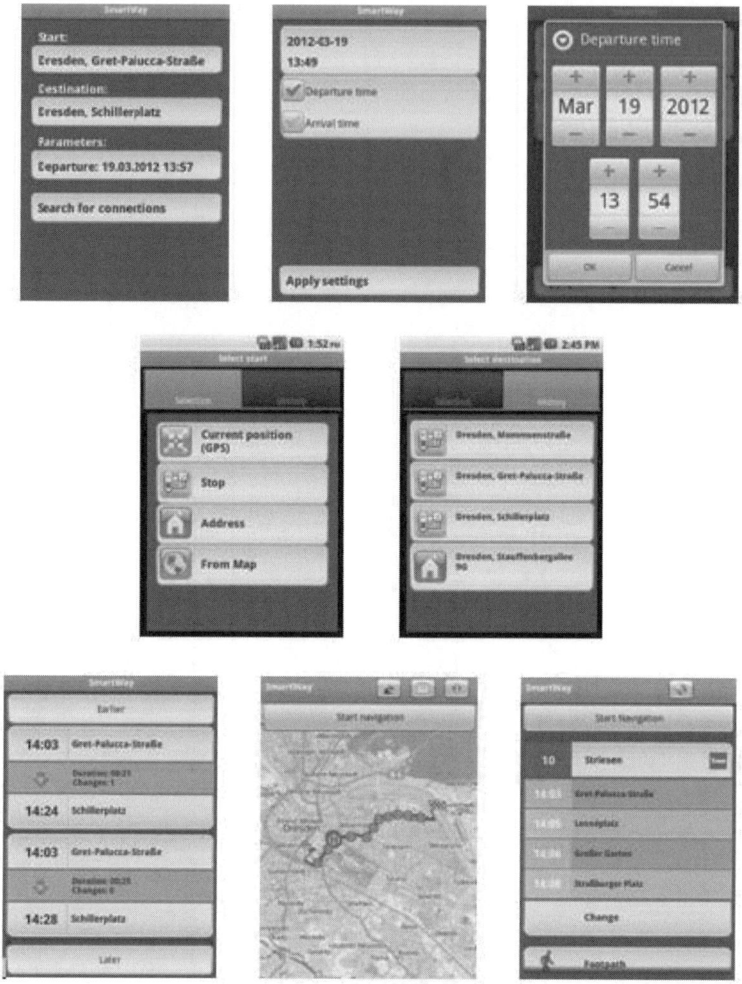

Notes: From top to bottom, screenshots on: get connection input masks; multiple ways to select the start and destination; select a connection; interchange at risk with connection update; better connection after early arrival; update the connection after deviation from the route.

Figure 3.1 Smart-Way features

phone application, and also one that may be installed on very basic mobile telephones so that a smartphone is not necessary. This could be a problem because all current applications work only on up-to-date smartphones. In fact, during the test of the Smart-Way application with both German and

Figure 3.1 (continued)

Italian users, the difficulty in using it with older releases of smartphones was clearly evident.

Another interesting result is that market segmentation seems to be the only viable solution for funding the service, which otherwise risks being expensive for the PT transport companies. The market should be divided into two large segments: people on repeated regular trips (work, study or other frequent trips) and people on leisure trips. The regular travellers should have an application offering real-time information to optimize their travel time, check traffic conditions in real time, assist with watching their children, and to support them in the event of emergency. For leisure travellers, time optimization is not the primary aim and the use of the device could be 'on demand', on a time basis (for example, for one day or one week), according to their holiday needs and location. The usage price must take into account that Smart-Way could be used abroad in places totally new to the users where it would be particularly useful to locate and guide them. Users are willing to buy a 'package' in which Smart-Way could be sold with a tourist guide and map. The price could vary with the time availability of the application: different expiry dates would influence the selling price. This implies a different business model, considering a different and, usually, higher WTP for an 'inessential' service, but allowing a high quality of leisure travel.

The declared WTP for real-time information (from 5 to 10 euros per month or 7–8 per cent of the monthly cell-phone expenditure) is consistent with the PROMISE project outcomes (Ojala, 2001) – which stated the primary role of cost in users' needs – and is comparable with the values in Dziekan and Kottenhoff (2007), who obtained a WTP in the range of 5–20 per cent of the ticket price for the trip, but also nothing at all, as also shown by Chorus et al. (2006).

Further segmentations should be considered for disabled and elderly users. Disabled users have specific needs regarding the information required and the usability of the device; current smartphones may not be suitable for them, while a more traditional mobile phone could work better. They also have higher willingness to pay than other people, because of the added value they may get from the device. They declared that such devices (incorporating the features presented in section 3.1) could radically change their quality of life, allowing them to travel without any setbacks thanks to the real-time knowledge of travel conditions. Disabled people like to travel, including for tourism, but they avoid doing it if they do not know the conditions they will face while travelling. The freedom of travelling has a huge value for them, bringing them closer to their social circle and other PT users.

Elderly people are a different case altogether since they have no willingness to pay, and see no real advantage in such an ICT for transport application. Thus, a social policy could possibly be devised to offer the application almost free of charge (social fare) as a bonus to the service for certain types of users such as the elderly (and, perhaps, disabled persons, despite their higher WTP) to compensate them for the problems they encounter with an offer often not directly addressing their needs.

In brief, the approach to follow with systematic users and the disabled or the elderly could be to add a price pro rata to the mobile phone subscription, or a variable price depending on the monthly mobile phone expense or pro rata to the PT pass for public transport users.

Within this framework, the transport companies have expressed their interest in contributing, although this application is now seen mostly as a back-up tool to collect data currently gathered using GPS.

Companies see Smart-Way as a tool to support their own data collection and to give users travel information and directions to stops. PT companies are willing to supply all the information useful for travellers in order to improve the use of public transport services. However, what a project like Smart-Way is going to do with these data remains for them an open issue. Besides, in some cities more operators compete in the public transport service market, in which case the operators do not intend to make their data available to competitors.

Using Smart-Way, PT companies expect to increase their customer volume, and obtain valuable statistics to plan and improve their services, confirming the findings of the Transit Cooperative Research Program (2003). The evidence that real-time information will generate sufficient demand to offset its costs is key for them, even though such services have become a clear trend and PT companies are becoming increasingly willing to spend on them (Lyons and Harman, 2002). The system also needs to be good enough to replace their current information system, since they now pay external providers to run it. Smart-Way has to be complementary to other contemporary solutions since the European Commission is working on a platform for a European ITS and Smart-Way must be compatible with it. Thus, the crucial point is that companies need standards for the system. Currently every application works with different data standards and languages, and every standard is a significant cost for the operators. This also fits well with users' needs, as discussed in this chapter.

5 CONCLUSION

The conclusion is that there is a market for Smart-Way, but serious attention must be paid to the features on offer and the business model. Both are focal to guarantee market penetration. A smart and clever design of the concept behind Smart-Way could convince transport companies to fully switch to this approach as they have already declared strong interest.

The research is continuing through the evolution of Smart-Way to a new application giving multimodal real-time information, whose need was already declared by Chorus et al. (2006). The new system was tested during the autumn of 2013 in the city of Lyon and the tests will continue in five other European cities: Turin, Madrid, Birmingham, Gothenburg and

Wroclaw. The research approach will be the same as for Smart-Way, but more structured and with a larger sample. The behaviour of the sample (150 people in each city) will be analysed before and after the use of the application (four months of tests) and monitoring will extend to all the transport modes (PT, car, bike, walk, car sharing), to provide an ICT application addressing integrated transport.

Early results about user needs in Lyon (with a sample of 50 individuals) before the experimentation, have confirmed the outcomes discussed in this chapter and are consistent with the literature. In fact, other authors have reported that people most inclined to use the ATIS instead of their own car (Zhang and Levinson, 2008; Grotenhuis et al., 2007; Williams et al., 2008) have a high educational level (Williams et al., 2008), have an open approach to technology (Neuherz et al., 2000), are familiar with using ICT to travel (Abdalla and Abdel-Aty, 2006; Khattak et al., 2003), and do not belong to the age group characterized as elderly (Zhang and Levinson, 2008). Furthermore, barriers possibly hampering the use of ATIS in encouraging travel behaviour change are similar to those discussed in this chapter, which can be summarized as the level of comfort and frequency of PT, the level of confidence with the technology, the lack of interest in the application, and its cost. Notably, the importance of usage cost of the application has clearly emerged from the Italian and German samples and has been confirmed by the larger French sample. Thus, a business model sustaining the application while keeping users' cost as low as possible is of the utmost importance. However, the improvement of PT services and soft interventions (mainly bike paths and parking) are also crucial to achieve modal diversion.

User needs discussed in this chapter may also be employed to identify opportunities and threats for the deployment of ATISs and, generally, ICT for transport. Opportunities include the potential increase of PT patronage and the interest in integrated transport information (as a tool for integrated transport policies). A significant opportunity concerns enabling disabled people to travel more confidently by PT, or to feel able to travel altogether, for example in new places, thereby improving their quality of life. Threats comprise the cost for users, as stressed previously, due to the low WTP reported. Threats are also related to the simplicity and ergonomics of the device, which is crucial for those not confident with the technology, as well as privacy matters, because some users are concerned about having their location continuously tracked.

Threats for companies and authorities include the current lack of standards and the related costs. Confidentiality of data, when operators compete with each other, is also a threat to the operation of ICT for transport.

A suggestion to decision makers is to run public campaigns to make ICT for transport more familiar to users, employing the outcomes of this research and using the same method for a European-wide bottom-up approach to identify both common and local user requirements.

ACKNOWLEDGEMENTS

This research has been carried out within the project 'SMART-WAY. Galileo based navigation in public transport systems with passenger interaction', funded by the European Commission under the Seventh Framework Programme, GALILEO.2008.1.7.1: LBS.

REFERENCES

Aarts, H., B. Verplanken and A. Van Knippenberg (1998), Predicting behaviour formations in the past: Repeated decision making or a matter of habit?, *Journal of Applied Social Psychology*, **28** (15), 1355–74.

Abdalla, F. and M.A. Abdel-Aty (2006), Modeling travel time under ATIS using mixed linear models, *Transportation*, **33**, 63–82.

Abdel-Aty, M. (2001), Using ordered probit modeling to study the effect of ATIS on transit ridership, *Transportation Research Part C: Emerging Technologies*, **9** (4), 265–77.

Abdel-Aty, M. (2002), *Design and Development of a Computer Simulation Experiment to Support Mode/Route Choice Modeling in the Presence of ATIS*, Orlando, FL: Civil and Environmental Engineering Department, University of Central Florida.

Abdel-Aty, M.A., R. Kitamura and P.P. Jovanis (1996), Investigating effect of advanced traveler information on commuter tendency to use transit, *Transportation Research Record*, 1550/1996.

Asakura, Y., E. Hato, M. Kashiwadani and S. Katsuki (2000), The simulation study of traffic information strategies using the computer for data collection on drivers' responses to ATIS, proceedings of the Fifth International Conference on Urban Transport and the Environment for the 21st Century, Southampton: WIT Press, pp. 343–52.

Bunch, J., C. Burnier, E. Greer, G. Hatcher, A. Jacobi, F. Kabir, C. Lowrance, M. Mercer and K. Wochinger (2011), Intelligent transportation systems benefits, costs, deployment, and lessons learned desk reference: 2011 update, Final Report, September 2011. FHWA-JPO-11-140, Washington, DC: ITS Joint Program Office, Research and Innovative Technology Administration, US Department of Transportation.

Chorus, C., E.E. Molin and B. Van Wee (2006), Use and effects of Advanced Traveller Information Services (ATIS): a review of the literature, *Transport Reviews*, **26** (2), 127–49.

Demographie konkret on line (2013), available at: http://www.demographiekonkret.de/Dresden_Wachsen_und_Schrumpfen.72.0.html (accessed 20 December 2013).
Dziekan, K. and K. Kottenhoff (2007), Dynamic at-stop real-time information displays for public transport: effects on customers, *Transportation Research Part A*, **41** (4), 489–501.
EC – European Commission (2008), *Action Plan for the Deployment of Intelligent Transport Systems in Europe*, COM(2008)886, Brussels.
EPOMM (2013), available at: http://www.epomm.eu/tems/index.phtml (accessed 20 December 2013).
European Parliament and Council (2010), *Directive 2010/40/EU on the Framework for the Deployment of Intelligent Transport Systems in the Field of Road Transport and for Interfaces with other Modes of Transport*, Brussels.
Farag, S. and G. Lyons (2012), To use or not to use? An empirical study of pre-trip public transport information for business and leisure trips and comparison with car travel, *Transport Policy*, **20**, 82–92.
Fujii, S. and T. Gärling (2003), Application of attitude theory for improved predictive accuracy of Stated Preference method in travel demand analysis, *Transportation Research Part A*, **37**, 389–402.
Gärling, T. and L. Steg (2007), *Threats from car traffic to the quality of urban life*, Amsterdam: Elsevier.
Gotzenbrucker, G. and M. Kohl (2011), Sustainable future mobility by ICTs. The impacts of Advance Traveler Information Systems on mobility behavior, *Proceedings of the 8th ITS European Congress*, pp. 1–15.
Grotenhuis, J., B. Wiegmans and P. Rietveld (2007), The desired quality of integrated multi-modal travel information in public transport: customer needs for time and effort savings, *Transport Policy*, **14**, 27–38.
Harris, P. and C. Konheim (1995), Public interest in, and willingness to pay for, enhanced traveler information as provided by IVHS in the New York Metropolitan Area, *Proceedings of the 5th Annual Meeting of ITS America, Vol. 1*, pp. 247–51.
HongCheng, G. and S. LiJun (2006), Advanced Traveler Information System for metropolitan expressways in Shanghai, Transportation research record, *Journal of the Transportation Research Board*, pp. 35–40.
Hyejung, H. (2009), *Measuring the Effectiveness of Advanced Traveler Information Systems (ATIS)*, Raleigh, NC: North Carolina State University.
Isfort (2011), La domanda di mobilità degli italiani, Rapporto congiunturale di metà anno, Rome: Audimob, Osservatorio sui comportamenti di mobilità degli italiani.
ISTAT (2012), Dati statistici sulla Provincia di Torino (Piemonte), available at: http://www.comuni-italiani.it/001/statistiche/ (accessed 20 December 2013).
ISTAT (2013), Annuario statistico italiano 2013, Rome: ISTAT.
Jou, R.C. (2001), Modeling the impact of pre-trip information on commuter departure time and route choice, *Transportation Research Part B*, **35** (10), 887–90.
Khattak, A.J., J.L. Schofer and F.S. Koppelman (1993), Commuters' en-route diversion and return decisions: analysis and implications for advanced traveler information systems, *Transport Research Board Part A*, **27** (2), 101–11.
Khattak, A.J., Y. Yim and L.S. Prokopy (2003), Willingness to pay for travel information, *Transportation Research Part C*, **11** (2), 137–59.

Krueger, R.A. and M.A. Casey (2000), *Focus Groups: A Practical Guide for Applied Research*, 3rd edn, London: Sage Publications.
Kumar, P., R. Dhanunjaya and S. Varun (2003), Intelligent transport system using GIS, *Proceedings of the Map India International Conference on GIS, GPS, Aerial Photography, and Remote Sensing*, New Delhi.
Lyons, G. and R. Harman (2002), The UK public transport industry and provision of multi-modal traveller information, *International Journal of Transport Management*, **1** (1), 1–13.
Mackett, R. and M. Edwards (1998), The impact of new urban public transport systems: will the expectations be met?, *Transportation Research Part A*, **32** (4), 231–45.
Neuherz, M., V. Patz, T. Pischner and H. Keller (2000), User acceptance and impacts of new multimodal traffic information services in BayernInfo, Working Paper.
Ojala, T. (2001), PROMISE – Project Summary, available at: http://www.cordis.lu/telematics/tap_transport/research/projects/promise.html, (accessed 1 August 2012).
Pronello, C. and C. Camusso (2011), Travellers' profiles definition using statistical multivariate analysis of attitudinal variables, *Journal of Transport Geography*, **19** (6), 1294–308.
Schweiger, C.L. (2003), Real-time bus arrival information systems. TCRP Synthesis of Transit Practice No. 48, *Transportation Research Board*, available at: http://onlinepubs.trb.org/onlinepubs/tcrp/tcrp_syn_48.pdf (accessed 23 June 2012).
Tang, L. and P. Thakuriah (2012), Ridership effects of real-time bus information system: a case study in the City of Chicago, *Transportation Research Part C*, **22**, 146–61.
Transit Cooperative Research Program (2003), *Real-Time Bus Arrival Information Systems: A Synthesis of Transit Practice*, Washington, DC: Transportation Research Board.
Transportation Research Group (2000), *Establishing User Requirements from Traveller Information Systems*, Final Report to Engineering and Physical Sciences Research Council.
Williams, B.M., H. Hu, A.J. Khattak, N.M. Rouphail and X. Pan (2008), *Effectiveness of Traveler Information Tools*, Final report, Raleigh, NC: North Carolina State University.
Zhang, L. and D. Levinson (2008), Determinants of route choice and value of traveler information: a field experiment, Transportation Research Record, 2086, 81–92.
ZVV (2013), Zurich Transport Network, available at: http://www.zvv.ch/en/timetables/mobile_timetable/iphone_application.html (accessed 20 December 2013).

4. The use of ICT to provide incentives to change habitual travel patterns*

Bernard Gyergyay

INTRODUCTION

Urban areas suffer from demand peaks due to the current set-up of their mobility systems. These urban demand peaks involve congestion on the road network and overcrowding on the public transport network that can result in negative health impacts due to air pollution and/or lack of exercise. Traditional solutions, such as building new roads or increasing public transport capacity, can be unsustainable and may come at a very high cost. In London, UK, these issues are also being addressed by promoting the most basic form of transport: walking. However, encouraging people to walk can be challenging, as people often prefer the comfort of their car or public transport even for very short trips.

Would people walk more if they are given an incentive to do so? This chapter presents a case study from London where incentives were used to encourage school students to walk instead of taking the bus. Information Communication Technology (ICT) was used in this case to record the distance students walked and to offer rewards based on how often they had walked to school.

This chapter showcases how ICT can be practically used as a vehicle to manage individual demand of urban transport systems more effectively, namely reducing peak time demand on buses. In this case study ICT is used as a cost-effective tool enabling the recording of users' desired travel behaviour causing urban peaks, allowing the distribution of rewards to them for using the suggested travel modes during peak times.

* The author of this work was employed by 'Transport for London' as a Transport Planner when this research was conducted (December 2009 to September 2010). Views about Step2Get are those of the author and do not reflect the opinion of Transport for London.

BACKGROUND

Walking to School – Why is it Important?

There are numerous benefits of walking to school.[1] A detailed overview by Tudor-Locke et al. (2001) highlights the importance of establishing an early habit of physical activity during childhood that is sustained into adulthood, and the importance of combating obesity worldwide. Increasing physical activity could be one of the key factors to prevent the prediction from coming true that by 2050, 40 per cent of adults in Europe will be clinically obese (International Obesity Taskforce, 2006).

However, in contrast to this notion, there has been an increase in the reported use of motorized vehicles. For example, in the UK, the proportion of trips to school made by car increased from 38 per cent in 1996 to 41 per cent in 2006 (Department for Transport, 2008). This chauffeuring of school children (age 3–18) reduces their physical activity and creates other problems, for example, road congestion during the morning peak, which has been labelled the 'school run'.

The above issues as well as a few others led to the creation of the Education and Inspections Act 2006 (clause 76, part 6) which places a general duty on local authorities in the UK to promote sustainable travel for journeys to, from and between schools and other educational institutions. The Act defines sustainable modes of travel as those that the local authority considers may improve the physical well-being of those who use them, the environmental well-being of all or part of the authority's area, or a combination of the two. Local authorities produce Sustainable Modes of Travel Strategies (SMoTS) to help fulfil their statutory obligations. Under this framework and as a reaction to the reducing number of walking trips, the UK government has created a statutory requirement for schools to implement a school travel plan.

Travel planning at school level is a strategic approach to more active, safe and sustainable journeys for education which help to increase the well-being and independence of children. The development of a school travel plan is based on consultation work with teachers, parents, pupils, governors and the wider community. A good travel plan should contain all the information about pupils' travel and transport needs and issues, such as details about all journeys to and from school, journeys to pre- and after-school events and journeys made during the school day to activities

---

1 Please visit http://www.walktoschool.org.uk for further information on walking to school.

at other locations (Transport for London, 2010a). Cairns et al. (2008) evaluate the traffic reduction potential of 'soft measures' like school travel plans, concluding that such approaches can deliver cost-effective outcomes, reducing overall congestion in urban areas.

In the context of London where this case study is set, the 'School Travel Plan Programme – Annual Report 2008/2009' and the School and Young People Delivery Plan published by Transport for London[2] (2010a and 2013) present an overview of the activities that are taking place to reduce the number of car trips on journeys to and from schools in London. The report highlights that between the hours of 8.30 am and 9.00 am, almost 20 per cent of all car journeys on the road include a stopover at a school, which contributes to congestion during this peak time. However, the programme does not discourage the use of public transport for the journey to school.

To encourage the use of public transport for the journey to school, young people until the age of 18 are able to travel free in London if enrolled in full-time education. However, this policy significantly increases bus demand and puts pressure on current bus capacity during peak times. London Buses[3] (TfL, 2009) estimate the cost of an additional morning peak bus to be £70 000 per bus, per year. However, this solution would further increase bus congestion in bus lanes and at bus stops, referred to as 'bus on bus congestion', resulting in delays. Walking for short journeys can reduce demand for bus services and thus decrease costs for providing local transport services. However, TfL (2008) has conducted research which highlights that walking is not popular and is mutually discouraged among young people (12–18 years of age), which raises the question how this reluctance to walking can be overcome.

In the Public Health White Paper 'Healthy Lives, Healthy People' (Department of Health, 2010), the UK government emphasized the use of incentives and 'nudging' to encourage positive, healthy behaviour change. Nudging has underlined the value of combining approaches that can simultaneously affect our capability, opportunity and motivation to change behaviour, including the use of incentives (Hunter et al., 2012). There is an ongoing debate about the use of travel behaviour incentives, for example reward systems that can alleviate congestion, and such

---

2 Transport for London (henceforth: TfL) is the integrated regional transport authority for Greater London. TfL manages the London Underground, Transport for London Road Network (580 kilometres of road), London Buses and London Rail. For further information see: www.tfl.gov.uk.

3 London Buses is the subsidiary of Transport for London that manages bus services within Greater London. All services are provided by private operators.

incentives seem to be particularly suited where tolls or fare increases are unacceptable for political reasons or unfeasible due to technical difficulties (Knockaert et al., 2012; Kusumastuti et al., 2011; Rouwendal et al., 2012).

Case Study Context: Wimbledon, London

This chapter specifically describes a case study in Wimbledon town centre, in the outer London Borough of Merton (2011 population: 20 500). Wimbledon Station is a major regional transport interchange where underground lines, trains, trams and buses meet. It is one of the busiest stations in outer London as in 2010–11 approximately 30 million passengers used the rail-based services London Underground, National Rail and Tramlink.

During the morning peak (07.50 am to 08:20 am) at Wimbledon Station hundreds of school children congregate and socialize on their way to school, resulting in severe overcrowding on buses and at bus stops. This causes operational issues for London Buses and intimidates other passengers and passers-by who have often complained to the transport authority and the police (TfL, 2009).

A solution was needed to reduce the peak bus demand of school children and bus stop overcrowding. A conventional solution would have been to increase the number of buses on the route to meet the demand, but this would have increased bus operation costs and 'bus on bus' congestion. More officers of the Safer Transport Police,[4] a service funded by TfL, were deployed on site to manage the number of people getting onto the bus and to reduce crowding at the bus stop, but the problem persisted.

A solution was needed where demand is shifted away from the overcrowded bus stop during the morning peak. Encouraging school children, who are the main cause of the overcrowding, not to take the bus but to walk for a short distance instead would reduce crowding and thus supersede the need for additional buses. The urban form and spatial surroundings have been considered when reviewing different options. This led to the creation of the Step2Get scheme which was designed: 'to demonstrate that by encouraging children to walk to school, the problems in Wimbledon town centre and on buses can be alleviated without increasing the capacity of the bus services or increasing the levels of policing' (TfL, 2009). Note that the above scheme objectives entail safety improvements at the overcrowded bus stops, and also invite an increase in physical activity for

[4] Each London Borough has a Safer Transport Team that is dedicated to policing and tackling crime on the transport network and engaging with the travelling public. The teams patrol buses and at transport interchange points, for example bus stations.

school children, which correspond with wider policy objectives as outlined previously (TfL, 2014 and UK Active, 2014).

Ricards Lodge all-girls secondary school was identified as a suitable partner in this project. With 1189 students ranging between the ages of 11 and 16 (Year 7 – Year 11) it was assumed that they could have a positive impact on reducing bus demand and overcrowding at the bus stop, especially because the total bus journey (Figure 4.2: Route A) time (including waiting, travelling and walking to school) from Wimbledon Station is 21 minutes[5] compared to 16 minutes[6] walking (Figure 4.2: Route B) time for the same journey. Students who did not travel through Wimbledon town centre on their way to school did not qualify for the scheme as they did not contribute to overcrowding at the bus stop.

Considering the shorter journey time in a generally safe walking environment compared to taking the bus, it is clear that walking should be the preferred mode of transport to Ricards Lodge School. On the contrary, students appeared to habitually prefer taking the bus because of reasons other than journey time, which calls for schemes that increase the value of walking over and above saving time. Consistent with this idea, focus groups with Ricards Lodge students established the fact that the student population would be more likely to walk to school if some kind of incentives were offered (TfL, 2009). This inspired the development of the ICT-based Step2Get travel behaviour change scheme outlined in the following section.

Case Study Context: the Step2Get Scheme

TfL worked in collaboration with the company 'Intelligent Health' to set up the walking incentive scheme Step2Get. Following their parents' consent, Ricards Lodge students received a unique radio-frequency identification (RFID) card. Students in London are generally familiar with this or other ICTs, because similar cards are used as electronic public transport tickets. Arising privacy and surveillance issues are discussed in Chapter 9 in this volume. The students were instructed to swipe their card at 'swipe points' (small, robust boxes attached to existing sign posts) along a walking route to Ricards Lodge School. A TfL (2009) route safety audit found no concerns with the walking route, an important precondition for setting up such schemes. These 'swipe point' boxes (Figure 4.1) contain affordable off-the-shelf mobile phones that feature integrated card reader

[5] Journey time was calculated using timetable information from the TfL website. (This does not factor in the likelihood of bus delay on days with high levels of congestion.)

[6] Estimated walking times at an average speed of 4.8 km/h.

Figure 4.1 Examples of swipe points in Wimbledon

technology which registers individual students when they swipe their RFID card. The walking proof recorded in this way is sent to a central database via a standard mobile phone short message service. The swipe points were only active on the way to school.

The only way of earning rewards without walking would be to give one's RFID card to another student to swipe. In order to prevent such fraud, the third swipe point was positioned within the school grounds, where school staff checked for and thus prevented multiple swipes performed by single students. This showcases the need for manual control of this ICT scheme to significantly increase accuracy and reliability. The Step2Get scheme was launched on 20 April 2009, immediately after the Easter holidays, and was stopped on 14 June 2010. Figure 4.2 shows the walking route with the exact location of the three swipe points.

For every completed walk, the students were then awarded a point, which was displayed on their individual Step2Get web account. Participating students could monitor the number of days that they walked the route and how many more days they need to walk before earning a reward. A cinema

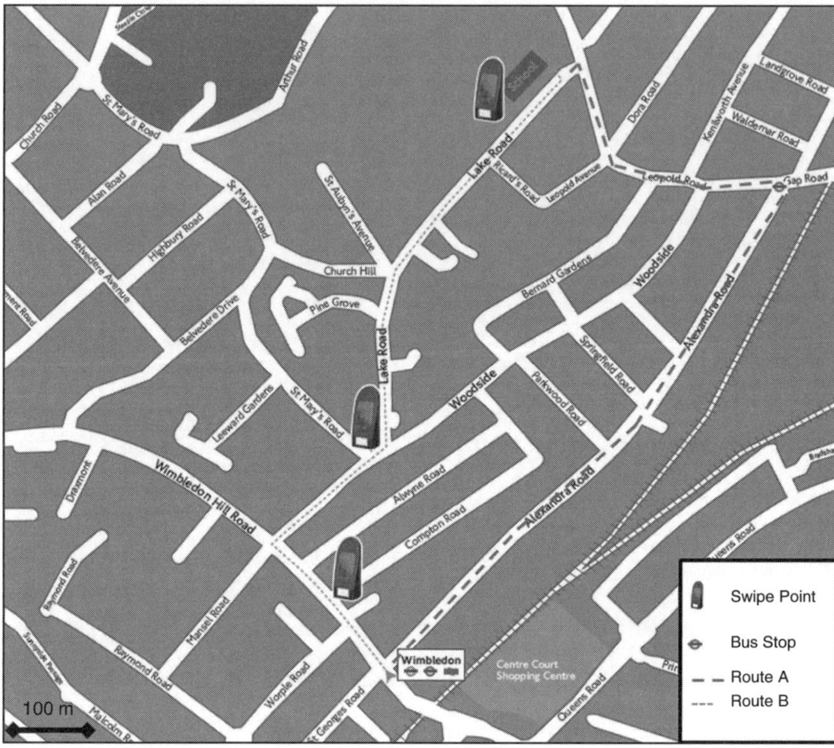

Figure 4.2 Map of the walking route between Wimbledon town centre and Ricards Lodge School and swipe point locations

ticket and a £5 Topshop[7] voucher could be claimed after completing five and eight 'walks', respectively. An automated email was sent from the web-portal (Figure 4.3) to inform the school as to which students had earned and redeemed rewards. The student then collected the reward from the member of school staff in charge of administering the scheme.

Generally speaking, the advantages of using an ICT-based, compared to a paper-based, incentives scheme are the following:

- ICT enables informative data to be efficiently recorded above and beyond just recording walks;
- ICT can inform walking time, walking speed and group walking analyses by analysing the exact time participants swipe their cards;

[7] A popular fashion high street shop in the UK.

Figure 4.3 Step2Get web-portal

- the use of ICT results in a considerable reduction of operational costs connected to walking data collection and data entry;
- ICT enables individual and group walking behaviour to be checked and compared on the web, and this can be done in real-time and in a user-friendly way.

SCHEME IMPACTS

Impact on Overcrowding and Bus Operations

Surveys carried out by TfL evaluated the impact of the Step2Get scheme on overcrowding and bus operations at Wimbledon Station. During the Step2Get scheme the average morning peak (07:50 am to 08:20 am) bus dwell time[8] fell by 62 per cent. Before Step2Get was implemented, bus drivers were forced to close bus doors before all waiting passengers were able to board. During the scheme, the number of passengers unable to board morning peak buses fell by 50 per cent. Additionally, there was a 48 per cent reduction in Safer Transport Police time required to deal with the overcrowding issues, such as operational issues for London Buses, stress for bus drivers and intimidation for passers-by. In sum, the Step2Get

[8] The scheduled time a bus is allowed to discharge and take on passengers at a stop, including opening and closing doors.

Table 4.1 Cost/benefit ratio for Step2Get stakeholders

Stakeholder	Average cost avoided (£ p.a.)	Percentage of total costs avoided	Cost (£ per annum)	Cost/benefit ratio
TfL	127 500	52	5200	**1: 24.5**
Bus Operators	1128	0.5	50	
Local Authority	106 666	43.5	4350	
Police	10 176	4	400	
All stakeholders	245 470	100	10 000	

Source: TfL (2010b).

walking incentive scheme demonstrates that overcrowding at public transport stops can be reduced by encouraging students to walk to their school.

The 2009 TfL evaluation suggests that the average cost of future implementations of the Step2Get schemes throughout London are approximately £10K per annum per school, with the walking rewards forming the main component of that cost. If all stakeholders were to contribute to funding a year of Step2Get, the £10K required could be divided as shown in Table 4.1. Given the prediction of avoided costs (reduced bus demand during pre- and post-school peaks; cost of bus-related anti-social behaviour; cost of on-street crowd management) the resulting overall cost/benefit ratio is 1:24.5. If these predictions are correct then future Step2Get schemes are worthy of funding even if assessed purely on financial grounds.

An additional, not quantified benefit of the Step2Get system was that it was programmed to discount swipe attempts after morning classes had begun, which encouraged participants to arrive at school on time. As a consequence, Ricards Lodge School reported a reduction in the number of students per day arriving late.

Impact on Habitual Travel Behaviour

The previous section suggests that incentives, such as the cinema tickets and shopping vouchers offered in Step2Get, can increase the frequency with which students walk to school. But what happens once the rewards stop? Do the incentivized students continue walking, or do they revert to taking the bus? The theory of habitual travel patterns (Gaerling and Garvill, 1993; Ronis et al., 1989) offers an explanation for how the incentives may have a lasting impact on the travel behaviour of the participating students.

According to this theory, a habit is defined as a behaviour sequence that is repeatedly performed without deliberation (Gaerling and Garvill, 1993; Ronis et al., 1989; Verplanken et al., 1997), which means that a previously formed behaviour script is automatically activated and retrieved from memory in an effort to respond to a familiar task in a time- and energy-efficient fashion (Gaerling and Axhausen, 2003). The process towards developing such a behaviour script involves:

1. a novel task (e.g. journey to a new school) for which there is no previously scripted behaviour;
2. the realization and pursuit of a preferred outcome (e.g. arriving at school in an efficient way);
3. deliberate consideration and comparison of known behaviour options with which the preferred outcome may be achieved (e.g. checking possible routes using an online journey planner);
4. choosing the subjectively superior option;
5. a process during which this particular choice is repeatedly made with increasingly less effort until no deliberation is needed any more (see Figure 4.4).

At this point, the subjectively superior behaviour option has transformed into an automatically activated and applied behaviour script (for

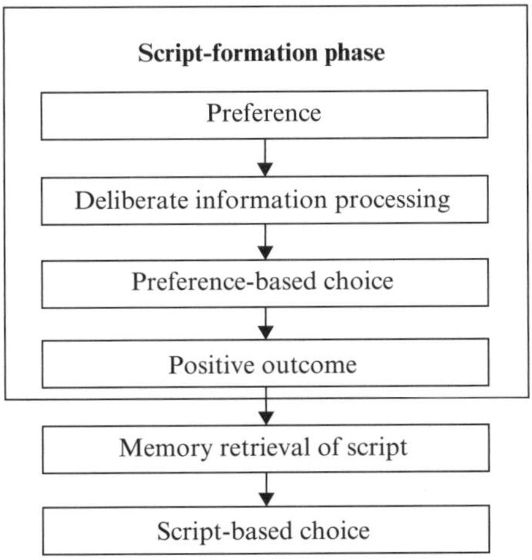

Source: Gaerling et al. (2001).

Figure 4.4 The transition of deliberate preference-based to script-based choice

example, the daily commute), which means other behaviour options won't be considered any more. Following this logic, a behaviour change intervention has to disturb and disrupt the automatic retrieval of an existing behaviour script (for example, taking the bus) by enforcing deliberation about an alternative behaviour (for example, walking).

To give some examples, the studies by Garvill et al. (2003) and Fujii and Kitamura (2003) investigate ways in which the car-use habit can be transformed into a habit of using public transport. Garvill et al. (2003) forced subjects to deliberate about possible means of transport by asking them to answer questions concerning their choice before they start home-based trip chains. The intention was to force subjects to deliberate. A temporary decrease of car use was observed. In the study by Fujii and Kitamura (2003) attitudes towards bus use became more positive and the frequency of bus use increased for at least as long as one month after the expiry of free bus tickets that the participants had received at the beginning of the study. At the same time car choice became less habitual.

The Step2Get scheme combines the enforcement of deliberation about an alternative behaviour choice with the promotion of incentives that try to increase the value of this alternative behaviour over and above the value of the current scripted automatic behaviour. More specifically, the intervention aimed at disturbing and disrupting the process of automatically retrieving the 'bus travel' script by means of advertising and incentivizing walking behaviour in an effort to replace the 'bus travel' script with a newly formed 'walking' script in the students' and their parents' perception.

To investigate whether Step2Get succeeded in doing so, groups of participating and non-participating students were surveyed with a nine-item questionnaire that assessed four aspects of habitual travel behaviour. First, the school journey modal share before and after the scheme ('Please tick the box that best describes your usual journey to school', response options: dropped off by car, taking the bus, walking, and other). Second, the frequency of walking to and from school before and after the scheme ('How often did you walk to/from to school in the morning each week from Wimbledon town centre?', response options: 0–5 times). Third, the change of walking behaviour on non-school trips ('Have you started to walk more to places other than school since the scheme began?', response options: more places, less places, same amount of places). Finally, fourth, attitudes towards walking to school before and after the scheme ('What do you think about walking to school?', written responses were coded as negative, neutral or positive attitude). The combination of these measures provides a comprehensive overview of any changes in the habitual travel pattern caused by the Step2Get scheme.

Due to the start of the summer holidays, the surveys were handed out to 1200 students two weeks after the end of the behaviour change scheme: 232 students had been willing to participate in Step2Get and 968 students did not participate; 79 Step2Get participants (response rate: 31.5 per cent) and 121 non-participants who could have qualified to participate in Step2Get returned their questionnaires.

The collected swipe card data was not considered, because there was no swipe card data from non-participants, and also because the participants did not provide swipe card data before the launch of the incentives and after the end of the incentives, which is why analyses of (the persistence of) changes in the pupils' habitual travel behaviour were only possible by means of a questionnaire. Future schemes could address this by supplying both participants and non-participants with ICTs before and after the scheme.

HABITUAL TRAVEL BEHAVIOUR – SURVEY RESULTS AND DISCUSSION

The first measure of habitual travel behaviour that was surveyed was the modal share of participants and non-participants before and after the Step2Get scheme. The survey results show a 10.39 percentage point increase in walking for participants and a 2.48 percentage point increase in walking for non-participants between the 'before' and 'after'. A Chi Square test was undertaken to establish whether these increments between the 'before' and 'after' are statistically significant. The results show that the increase in walking is statistically significant for participants ($p < .01$) but not for non-participants.

The second measure of habitual travel behaviour that was surveyed was the 'before' and 'after' frequency of walking to school and walking back after school. For both participants and non-participants there was an increase between the 'before' and 'after' in the number of students who walked to school five times a week. However, this increment was almost four times larger for the participating students (6.33 percentage points) compared to the non-participating students (1.65 percentage points). An analysis of the student population that never walked to school revealed an 8.86 percentage point reduction of that group for participants while the percentage of non-participants who never walked did not change at all. The survey results for walking from school, for which participants did not receive rewards, show that there was a 5.06 percentage point increase for participants and a 3.31 percentage point increase for non-participants in the frequency of walking five times a week from school.

The third measure of habitual travel behaviour that was surveyed was the change of walking behaviour on non-school journeys of participants and non-participants: 49.37 per cent of Step2Get users stated that they walk more often to non-school destinations compared to only 33.88 per cent of non-participants. A Chi Square test revealed that this difference between participants and non-participants is statistically significant ($p < .001$).

The fourth measure of habitual travel behaviour that was surveyed was attitude towards walking to school 'before' and 'after' the Step2Get scheme for participants and non-participants: 75.56 per cent of the participants and 20.66 per cent of the non-participants who previously had negative or neutral attitudes about walking to school developed a more positive attitude. A Chi Square test was run, which shows that this difference between the attitude change of the participants and the attitude change of the non-participants is statistically significant ($p < .05$). Furthermore, not a single participant changed from a positive or neutral attitude towards a negative attitude, whereas 18.29 per cent of non-participants developed a more negative attitude towards walking to school.

The survey results from the four measures of habitual travel behaviour suggest that the combination of enforcing deliberation about walking as an alternative travel behaviour and issuing temporary rewards for walking repeatedly may effectively disturb and disrupt the retrieval of the existing 'bus' script and promote the formation of a 'walking' script in student travel choice perception. This new 'walking' script has been shown to both persist despite the end of the rewards and to extend to journeys other than the one for which the incentives were issued in the first place.

The results for the modal share show that there is a significant increase in the number of participants who walk and that this increase persists even after the rewards have stopped. This suggests that a noticeable proportion of the participants have replaced their 'bus' script with a 'walking' script, presumably because the subjective value of walking has surpassed the subjective value of taking the bus. This argument is supported by the results for the second and third measure of habitual travel behaviour (Figure 4.4). The second measure suggests that participants have extended their recently formed walking script to journeys from school even though they did not receive any additional rewards for this journey in the first place. The results for the third measure show that after the end of Step2Get, participants are significantly more likely to consider walking in their script-based decision-making process for journeys to non-school destinations, which suggests that walking has become more of a norm. If this is true, then the participants should prefer to walk whenever confronted with a new, non-scripted, but walkable destination.

The results from the fourth measure show that positive attitudes towards walking to school were significantly more prevalent for participants after the scheme. This suggests that some students corrected their exaggeratedly negative attitude towards walking to school as a result of their experience of walking to school on a regular basis for the purpose of attaining some reward. A positive attitude towards walking to school is a fruitful breeding ground for the development and maintenance of an actual preference for walking that will influence current and future scripts.

The synopsis of the assessed travel behaviour measures enables a comprehensive evaluation of the impact of Step2Get on habitual travel patterns, and the observed results demonstrate that these habitual travel patterns have changed in the direction of the desired outcome, namely more walking. This finding highlights a clear benefit of ICT for transport used in such contexts.

Limitations of the Project Approach

Generally the scheme was very well received by the students, the parents, the school, the media and UK government (Martin, 2010; Donnelly, 2010; Cabinet Office, 2010). During this research study it became clear that there are some limitations in the Step2Get scheme applied at the Ricards Lodge School, and that these limitations should be addressed in similar schemes in the future.

While analysing the survey results it became apparent that many students who registered with Step2Get were already walking to school at that time. Future interventions should avoid wasting resources through rewards encouraging an already existing desirable behaviour. A similar challenge has been faced about peak time drivers in a Dutch policy experiment called 'Spitsmijden' (Avoiding the Peak), pointing out that existing positive travel behaviour should be acknowledged too (Knockaert et al., 2012). This discussion falls outside the scope of this chapter as it raised issues about equity which have been outlined elsewhere in the literature (Thomopoulos et al., 2009).

In the survey several students highlighted their concern, which is to some extent shared by the author too, that handing out rewards corrupts the decision making of students and does not foster an intrinsic desire to change travel behaviour. Moreover, handing out rewards may possibly undermine other travel behaviour change programmes, for example the TfL School Travel Programme, as students might expect a reward for any subsequent travel behaviour change. This is a wider concern raised by such schemes, particularly due to the students' age, so future projects can mitigate this concern by integrating Step2Get into the other wider education programmes that are running parallel to the intervention.

A significant amount of resources of the Step2Get intervention were spent on cinema and shopping vouchers that were handed out as tangible rewards to the students. Future projects should evaluate the option of using intangible rewards as a reward stimulus, such as game points. The process of 'gamification', the use of game design elements in non-game contexts, could further act as a motivation to change behaviour (Deterding et al., 2011). This is particularly relevant in the context of the budget constraints that many local authorities in Europe face after the 2008 economic crisis, which means that politically it can be extremely difficult to spend public funds on tangible rewards.

Another issue to be addressed is the parents' involvement in the scheme, as there were several cases where students were not allowed to participate in Step2Get due to parental safety concerns about walking. In future schemes incentives could be extended to include parents as well as students, whilst mitigating such safety concerns through 'walking buses' or other collective actions.

CONCLUSION

The Step2Get scheme was thoroughly evaluated with regards to its performance in the reduction of overcrowding, the improvement of bus operation and the impact on habitual travel patterns of participants. Limitations were also highlighted and should be addressed by future schemes. Nevertheless, the results of this pilot study are promising and open up opportunities for more similar ICT for transport applications while justifying future applications of similar types of behaviour change interventions that might target a range of population groups and journey destinations, such as universities or work places. Hunter et al. (2013) give an example of how the system can also be applied to other walking-related policy goals, such as increasing physical activity and health which is a contemporary policy target in the UK (UK Active, 2014), whereas Tight et al. (2011) offer an overview of relevant transport visions highlighting the importance of such schemes.

As highlighted in the introduction, urban areas suffer from road congestion, overcrowding on the public transport network, air pollution and health issues. The approach of Step2Get could present a suitable and cost-effective solution to tackle them and its core idea of shifting demand to environmentally sustainable and healthier modes of transport presents an alternative to increasing road or public transport capacity. Therefore this intervention should be included in the standard portfolio of travel demand management tools (for example, road pricing or work travel planning) that are applied by urban and transport planners to manage travel demand as an alternative to increasing capacity. In this regard, it should be noted that one of the major assets of the behaviour change intervention outlined and evaluated here is that it does not need to be applied permanently, despite being founded on the premise that it will create a permanent positive impact. However, this was not achieved during behaviour change intervention where incentives were given to car drivers to avoid morning rush hour driving on a designated corridor, documented in Knockaert et al. (2012) and Bliemer and Van Amelsfort (2010). The participants of 'Spitsmijden' quickly returned to their pre-intervention behaviour after rewards ceased (Ben-Elia and Ettema, 2011; Knockaert et al., 2012). When contrasting the post-intervention results of 'Spitsmijden' and Step2Get, it can be argued that participants of 'Spitsmijden' did not develop an intrinsic motivation to avoid morning rush hour driving whereas Step2Get participants did develop an intrinsic motivation to walk. This might be because there are more barriers for commuters shifting from the car than there are for students shifting from the bus.

The swipe cards and swipe points technology might appear archaic to the generation of 'digital natives' who are getting increasingly more used to navigating their everyday life with the help of an ever more functional ICT device: the smartphone. Websites, such as www.walkit.com, provide walking routes that can easily be accessed through a smartphone. Research shows that the use of smartphones often forms a habitual 'checking for updates' behaviour that essentially increases the use of all sorts of applications installed on the device (Oulasvirta et al., 2011). Future behaviour change schemes, such as Step2Get, might profit from the higher overall use of smartphone applications in a multitude of ways (for a review, see Holleis et al., 2012). In particular, smartphone applications can be programmed to automatically deliver feedback on achievements in the direction of desired behaviour change. This feedback has been suggested to motivate further behaviour change, and this effect can be boosted if the feedback is also shared in social networks. Despite the fact that GPS and other geolocation data are not 100 per cent accurate, as discussed in another chapter (Chapter 6), such functionality of smartphones can record a traveller's behaviour, and thus swipe card and swipe point technology presented in this chapter might become obsolete, which is both more user-friendly for participants and more cost-efficient for operators of the behaviour change schemes. However, there are data protection and surveillance concerns regarding GPS tracking and inequalities between students who have a smartphone and those who do not, all of which need to be addressed within such ICT for transport schemes in the future.

REFERENCES

Ben-Elia, E. and Ettema, D. (2011) Rewarding rush-hour avoidance: a study of commuters' travel behaviour. *Transportation Research Part A*, 45(2011), 567–82.

Bliemer, M.C.J. and Van Amelsfort, D.H. (2010) Rewarding instead of charging road users: a model case study investigation effects on traffic conditions. *European Transport/Trasporti Europei*, 44, 23–40.

Cabinet Office – Behavioural Insights Team (2010) Applying behavioural insight to health. London: Cabinet Office.

Cairns, S., Sloman, L., Newson, C., Anable, J., Kirkbride, A. and Goodwin, P. (2008) Smarter choices: assessing the potential to achieve traffic reduction using 'soft measures'. *Transport Reviews*, 28(5), 593–618.

Department for Transport (DfT) (2008) Travel to school personal travel factsheet. London: Department for Transport.

Department of Health (DoH) (2010) Healthy lives, healthy people: our strategy for public health in England. London: DoH.

Deterding, S., Dixon, D., Khaled, R. and Nacke, L. (2011) From game design elements to gamefulness: defining 'gamification'. Proceedings of the 15th International Academic MindTrek Conference: Envisioning Future Media Environments, pp. 9–15.

Donnelly, L. (2010) Children will be given shopping vouchers for walking to school. *The Telegraph*, 28 November. Available at: www.telegraph.co.uk (accessed 19 January 2014).

Fujii, S. and Kitamura, R. (2003) What does a one-month free bus ticket do to habitual drivers? An experimental analysis of habit and attitude change. *Transportation*, 30, 81–95.

Gaerling, T. and Axhausen, K. (2003) Introduction: habitual travel choice. *Transportation*, 30, 1–11.

Gaerling, T. and Garvill, J. (1993) Psychological explanations of participation in everyday activities. In: Gaerling, T. and Golledge, R.G. (eds), *Behaviour and Environment: Psychological and Geographical Approaches*. Amsterdam: Elsevier/North-Holland, pp. 270–97.

Gaerling, T., Fuji, S. and Boe, O. (2001) Empirical tests of a model of determinants of script-based driving choice. *Transportation Research Part F*, 4, 68–72.

Garvill, J., Marell, A. and Nordlund, A. (2003) Effects of increased awareness on choice of travel mode. *Transportation*, 30, 63–79.

Holleis, P., Luther, M., Broll, G., Hu, C., Koolwaajj, J., Peddemors, A., Ebben, P., Wibbels, M., Jacobs, K. and Raaphorst, S. (2012) TRIPZOOM: a system to motivate sustainable urban mobility. In *ThinkMind // SMART 2012, The First International Conference on Smart Systems, Devices and Technologies*, pp. 101–104.

Hunter, R.F., Davis, M., Tully, M.A. and Kee, F. (2012) The physical activity loyalty card scheme: development and application of a novel system for incentivizing behaviour change. *Lecture Notes of the Institute for Computer Sciences, Social Informatics and Telecommunication Engineering*, vol. 91, 2012, pp. 170–77.

Hunter, R.F., Tully, M.A., Davis, M., Stevenson, M. and Kee, F. (2013) Physical activity loyalty cards for behaviour change: a quasi-experimental study. *American Journal of Preventive Medicine*, 45(1), 56–63.

International Obesity Taskforce (2006) cited in Lobstein, T. and Millstone, E.

(2007) Context for ProGrow study: Europe's obesity crisis. *Obesity Reviews*, 8 (suppl. 2), 7–16.

Knockaert, J., Tseng, Y-Y., Verhoef, E.T. and Rouwendal, J. (2012) The Spitsmijden experiment: a reward to battle congestion. *Transport Policy*, 24, 260–72.

Kusumastuti, D., Hodgson, F., Thomopoulos, N., Hjamalrsson, A., Veenstra, S., van Berkum, E., Klok, E., Dethmers, V. and Groenewelt, B. (2011) D3.3 Impact of incentives, Sustainable Social Networking Services for Transport (SUNSET) Deliverable, FP7 no. 270228, Brussels: DG Information Society & Media.

Martin, D. (2010) Obese could be 'bribed' to walk with store vouchers and cinema tickets. *The Daily Mail*, 3 December. Available at: www.dailymail.co.uk (accessed 19 January 2014).

Oulasvirta, A., Rattenbury, T., Ma, L. and Raita, E. (2011) Habits make smartphone use more pervasive. *Personal and Ubiquitous Computing*, 16(1), 105–14.

Ronis, D.L., Yates, J.F. and Kirscht, J.P. (1989) Attitudes, decisions and habits as determinants of repeated behaviour. In Pratkanis, A.R. and Greenwald, A.G. (eds), *Attitude Structure and Function*, Hillsdale, NJ: Erlbaum, pp. 213–39.

Rouwendal, J., Verhoef, E.T. and Knockaert, J. (2012) Give or take? Rewards versus charges for a congested bottleneck. *Regional Science and Urban Economics*, 42, 166–76.

Thomopoulos, N., Grant-Muller, S. and Tight, M.R. (2009) Incorporating equity considerations in transport infrastructure evaluation: current practice and a proposed methodology. *Evaluation and Program Planning*, 32(4), 351–9.

Tight, M., Timms, P., Banister, D., Bowmaker, J., Copas, J., Day, A., Drinkwater, D., Givoni, M., Gühnemann, A., Lawler, M., Macmillen, J., Miles, A., Moore, N., Newton, R., Ngoduy, D., Ormerod, M., O'Sullivan, M. and Watling, D. (2011) Visions for a walking and cycling focussed urban transport system. *Journal of Transport Geography*, 19(6), 1580–89.

Transport for London (TfL) (2008) Understanding young people and active travel. London: Transport for London.

Transport for London (TfL) (2009) Wimbledon Project: evaluation report. London: Transport for London.

Transport for London (TfL) (2010a) School travel plan programme: annual report 2008/09. London: Mayor of London.

Transport for London (TfL) (2010b) 'Step2Get': London-wide roll-out feasibility report. London: Transport for London.

Transport for London (TfL) (2013) Transport for London school and young person delivery plan: setting our future direction (draft). London: Transport for London.

Transport for London (TfL) (2014) Improving the health of Londoners: transport action plan. London: Transport for London.

Tudor-Locke, C., Ainsworth, B.E. and Popkin, B.M. (2001) Active commuting to school: an overlooked source of children's physical activity. *Sports Medicine*, 31(5), 309–13.

UK Active (2014) Turning the tide of inactivity. London: UK Active.

Verplanken, B., Aarts, H. and van Kippenberg, A. (1997) Habit, information acquisition, and the process of making travel mode choices. *European Journal of Applied Social Psychology*, 27, 539–60.

PART II

ICT applications and sustainable transport: threats and opportunities

5. Spatiotemporal dynamics in public transport personal security perceptions: digital evidence from Mexico City's periphery

P. Christopher Zegras, Kuan Butts, Arturo Cadena and Daniel Palencia

1 INTRODUCTION

The potential for information and communication technologies (ICTs) to revolutionise transportation is of long-standing interest. Just as the telegraph, trans-oceanic cable communications, the telephone, and the fax surely influenced travel behaviours and supply of transportation infrastructures and services, related developments, especially in computing, have changed the way we analyse and plan systems. The most recent wave of ICT-related technological advances – epitomised by high-powered sensing and real-time computing capabilities – promises a new era of seamless services, autonomous vehicles, and high resolution system microsimulation, to name a few emerging opportunities.

In this chapter, we demonstrate the potential offered by one particular advanced ICT instrument, the smartphone, to improve our abilities to diagnose urban mobility problems. Our empirical setting is one of the world's great 'megacities', and our research topic is personal security, in particular users' perceived security reported on public transportation services. Our fundamental ICT-related aim is to test the value of the smartphone as a diagnostic tool to ultimately improve transport services, particularly in a megacity context. Our broader inquiry aims to understand what influences perceptions of personal security on public transport services in suburban Mexico City, with a specific focus on whether recently developed intermodal transfer stations and some of their related technological and design elements might play a role.

Technological Background: Smartphone-based Mobility Data Collection

Because of their ubiquity, portability and sensing capabilities, today's mobile phones can be a great source of data about individuals' mobility demand. Mobile phones' general ubiquity has made records from telecom companies one form of large-scale, though low-dimensional (for example, low spatial resolution, no user information), data source. Gonzalez et al. (2008), for example, analysed thousands of mobile phone user records and observed highly regular daily travel patterns, as revealed by the cell-phone tower locations of the phone's calling or texting activities. Today's so-called 'smartphones' – with advanced computing, networking (that is real-time data communications), and sensing (for example, GPS, accelerometers) capabilities and the ability to run programs (that is 'apps') and browse the web – offer a high-resolution means for purposeful, 'bottom up' data collection. Cottrill et al. (2013), for example, aim to replace traditional travel survey methods with a specialised app to non-intrusively (that is requiring minimal user interaction) collect location traces (using GPS, Wi-Fi, and cell tower locations) and accelerometer data; combined with a back-end, server-based intelligence the goal is to detect travel modes, stops, trip origins and destinations, and purposes. Jariyasunant et al. (2014) take a similar technological approach to Cottrill et al. (2013), but with the aim of providing feedback to the users (on, for example, their carbon footprint, calories burnt) to explore possible behavioural and attitudinal influences.

With small screens allowing data entry, smartphones also hold promise for improving contextual richness and multi-dimensionality of field data collected for other types of mobility analyses, such as on-board surveys and traffic counts. In theory, heterogeneous, high resolution field data can be rapidly collected, transmitted, fused and validated almost simultaneously. For example, passenger counts can be combined with on-board surveys, which can be precisely matched with time and place stamps and transmitted remotely to databases. In traditionally data-sparse settings, such as much of the global south, some of these capabilities have already been deployed to, for example, generate public service inventories in rural Nigeria (Berg et al., 2013). Special-purpose apps have also emerged to facilitate mapping public transportation services. TransitWand, for example, uses GPS signals to track a user's location and allows the user to collect stops, boarding/alighting times, and passenger counts. Users manually upload this information using Wi-Fi or cellphone networks; authorities in Mexico City recently used TransitWand to map semi-formal transit services in the city (Eros et al., 2014).

In January 2012, a research team at MIT in collaboration with a local organisation set out to test smartphone capabilities for understanding public transportation operations (mapping routes, speeds, stops) and conducting on-board surveys and passenger counts in Dhaka, Bangladesh. Subsequent improvements to the technologies and concentrated deployment of a volunteer team resulted in the development of the city's first widely available map of the city's semi-formal, privately operated public transport routes, released in paper and electronic form in May, 2013 (Zegras et al., 2015). We subsequently adapted the technology, described in more detail below, for testing in another of the world's megacities, Mexico City.

2 RESEARCH CONTEXT AND QUESTIONS

The Mexico City metropolitan area[1] is one of the world's largest megacities, with some 20 million people spread across at least 4000 km^2. The city's multimodal passenger transportation system includes one of the world's largest urban rail systems (Metro), an extensive, privately operated and loosely regulated system of minibuses (a range of vehicle types, generally referred to as *colectivos*), a five-line (as of 2014) bus rapid transit (BRT) system first introduced in 2005, and approximately 5.4 million private cars. As of the last household travel survey, in 2007, *colectivos* accounted for nearly 50 per cent of the city's motorised person trips (SETRAVI, 2010). *Colectivos* include *combis*, modified vans with inward-facing seating along the periphery of the internal cabin; and *peseros* (or minibuses), with larger capacity than the *combis*, and with front-facing seat rows.

Among the range of challenges facing the area's mobility system looms personal security concerns, or fear of crime.[2] Research suggests that fear of crime on public transport arises from perceived vulnerabilities, a result of factors over which passengers have little control (Yavuz and Welch, 2010). These factors might include vehicle and station appearance, service reliability, presence of security technologies and police, personal characteristics (e.g., age, gender) and experiences, and trip characteristics (Yavuz and Welch, 2010). In the UK in the 1990s, authorities estimated that fear of crime suppressed 3 per cent to 10 per cent of public transport demand (Newton, 2004). Gender plays an important role, with women's personal security needs and concerns long recognised as being different from men's (Lynch and Atkins, 1988; Smith, 2008). Women, for example, may fear

---

1 We use Mexico City to refer to the Metropolitan Area of Mexico City.

2 We use the word 'security' to refer to crime and fear of crime, not traffic safety (although traffic safety problems can result from criminal activities).

sexual harassment (for example, GDF, 2011; Hsu, 2011); men, on the other hand, may actually under-report their personal security concerns (for example, Sutton and Farrall, 2005) and may be more susceptible to police harassment/corruption.[3] Newton (2004) suggests the need to develop techniques to better analyse the linear patterns of public transport crime (and, by extension, fear of it), including internal and external environments of the routes, corridors, stops and stations; similarly, Smith (2008) advocates the 'whole journey approach' to analyse the 'dynamic and complex environment' (p. 129) influencing fear of crime.

While not widely studied in the academic literature, Mexico City's transportation-related personal security concerns are evident. As Davis (2007) states, the city's 'citizens have become so fearful of crime that they are changing where and how they live, travel, and interact with their fellow citizens' (p. 67); reportedly 48 per cent of residents avoid public transport due to fear (Davis, 2007). Women are particularly at risk; women comprised 95 per cent of the 850 victims of reported incidents on the Metro from 2008 to 2010 (GDF, 2011); the Metro has women-only cars and platform areas during rush hours and women-only buses were introduced in 2008 and by 2011 covered 50 bus routes carrying 25 000 passengers on an average work day (GDF, 2011).

CETRAMs: An Experimental Opportunity

Approximately 40 per cent of the metropolitan area's population resides in the Federal District (*Distrito Federal* or DF), the nation's capital and the historic economic, cultural and financial hub of the region and the country. The remainder of Mexico City's residents sprawl across a landscape of some 40+ municipalities in the surrounding State of Mexico. Functionally, this jurisdictionality is important since buses and *colectivos*, which provide the major share of public transportation services outside the DF, cannot legally operate in the DF. Thus, bus and *colectivo* users traveling from the broader metropolitan area into the DF must transfer at the border, typically at intermodal transfer centres, known as CETRAMs (*Centros de Transferencia Modal*).[4] As of mid-2013 (the time of the fieldwork presented below), 46 CETRAMs functioned in the DF, 39 of which interface with

---

[3] We could find no academic literature supporting this concern in Mexico City; however, our field work produced anecdotal evidence that men on *colectivos* were more likely to be subject to police 'shakedowns' (vehicles being stopped and men being pulled off by police and forced to pay 'fines').

[4] Note that not all CETRAMs are on the periphery; many function at intermediate Metro stations. In addition, not all CETRAMs are technically inter-modal, as some just provide transfers between different services of the same mode.

Metro and BRT stations. The DF's CETRAMs serve roughly five million passengers each day (approximately 40 per cent from outside the DF), making them the city's most-used public transport facilities (GDF, 2011).

Despite their intense utilisation, most CETRAMs remain little more than large parking lots. A recent report conducted for the DF (GDF, 2011), evaluating the design and implementation of public transport policy in the city, highlights a range of CETRAM problems including inadequate design and infrastructures, lack of maintenance, high levels of congestion, poor administration, disorganised informal commerce, and poor levels of public security, especially for women. The DF government strategy to improve CETRAM conditions and functioning includes public–private partnerships (PPPs) to transform CETRAMs into urban spaces that not only improve the passenger transfer experience, but offer a range of formal commerce and services, provide employment opportunities, improve the urban surroundings, and increase security (GDF, 2011).

We examine two pioneering CETRAMs, developed by private investors under 30-year PPP concessions: El Rosario, inaugurated in May 2013; and Mexipuerto (also known as *Ciudad Azteca*), inaugurated in November 2009. The same developer developed and operates, under separate concessions, both CETRAMs, with most income coming from rents from the various real estate activities, which include movie theatres, retail, and social services, including a medical clinic in the Mexipuerto case. Security measures employed include closed circuit television and private security personnel. Both provide intermodal links with Metro lines; Mexipuerto also integrates with a BRT terminal station.

Located on the northwest (El Rosario) and northeast (Mexipuerto) boundaries of the DF, the CETRAMs are situated in different jurisdictions: El Rosario in the DF *delegación*[5] of Azcapotzalco, and Mexipuerto in the State of Mexico *muncipio* of Ecatepec (Figure 5.1). This implies different public security authorities and different regulatory authorities and operating procedures within the CETRAMs. At El Rosario, for example, the DF manages vehicle activity within the terminal, whereas at Mexipuerto the concessionaire manages the intermodal services, for a fee to the vehicle operators (taxes, buses, *colectivos*) collected through a radio-frequency identification (RFID) based toll system.

The areas that the two CETRAMs serve are somewhat different, socio-economically and demographically (Table 5.1).[6] Located in Ecatepec, one

---

5 *Delegaciones* are sub-districts in the DF.

6 Information on the approximate CETRAM service areas was gathered by taking the point location of the on-board surveys conducted and linking that point to the census tract it contained.

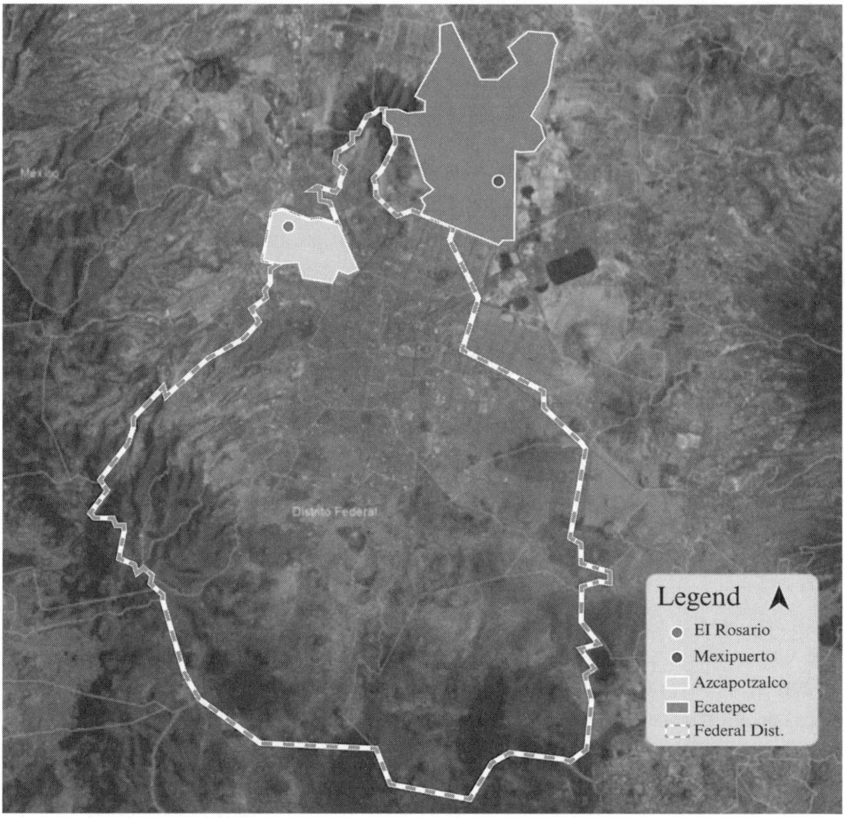

Figure 5.1 The locations of the two CETRAMs in the metropolitan area

of the oldest and largest suburban *municipios* of Mexico City, Mexipuerto serves a much larger population, although with lower density than El Rosario, with a higher share of children and lower share of older adults. The Mexipuerto service area has lower unemployment, but also lower household motorisation rates. In the immediate station area vicinities, El Rosario has become more service-oriented in the past decade, with a newly completed multi-storey mall directly adjacent and two large housing complexes, two technical high schools and a university campus nearby. Adjacent to Mexipuerto is 'Multiplaza Aragon', one of Mexico City's largest shopping centres, first opened in 1971, and anchor to Ecatepec's main commercial corridor.

The two CETRAMs have comparable amounts of space – approximately 70 000 m^2 each of real estate. Architecturally, some differences arise. El

Table 5.1 Socioeconomics and demographics along CETRAMs routes

Variable	El Rosario	Mexipuerto
Population	211430	418178
Population density (persons/hectare)	160.6	116.5
Share female	52.4%	51.6%
Population 0–14 years old	18.7%	27.3%
Population 15–64 years old	67.4%	67.0%
Population 65+ years old	9.2%	4.8%
Economically active population	43.7%	42.4%
Unemployment rate	6.1%	4.9%
Occupants/house	3.6	3.7
Vacant housing	14.3%	26.0%
Share households with motor vehicle	54.1%	36.5%
Percentage of population with at least 1 year of university education	25.0%	29.8%

Sources: INEGI (2010, 2000).

Rosario is more akin to a 'fortress in the air', with limited, gated entrances to the central structure, leading consumers to the second floor commercial areas and primarily accessible only by off-boarding bus commuters or Metro riders. Street access is limited via two back stairwells controlled by armed guards. Mexipuerto presents a more open façade to the immediate surroundings. As mentioned previously, the vehicle management schemes in the two CETRAMs also differ. El Rosario functions traditionally, with few improvements over the pre-concession site; the DF is responsible for management, from an on-site office and via random vehicle verification. In contrast, at Mexipuerto all vehicles entering the facility must carry an RFID chip, which the concessionaire uses to charge the transit operators for use of the facility as well as to organise vehicles within the facility (at passenger loading bays and so on) – a form of on-site intelligent transportation service (ITS). Both stations service roughly 10000 vehicles per day, with El Rosario handling over 180000 passengers per day and Mexipuerto handling about 110000 passengers per day.

Research Questions

In light of Mexico City's goals of enhancing CETRAMs to improve public transport system services, the surrounding environment, the user experience and public security, we aim to examine these new, enhanced CETRAMs with a particular focus on personal security. We attempt to measure individuals' perceived security as self-reported in a survey, aiming to represent fear of physical violence, crime or other social intimidation (Davis, 2013). In the Mexico City context, the CETRAMs offer a relevant lens on the security question, representing the private sector use of real estate, security forces and technological surveillance, which have been increasingly deployed towards security enhancements in the city (for example, Davis, 2007; 2013). In particular, we aim to answer the following questions:

- Are the new CETRAMs increasing the sense of security among public transport users? If so, does this effect endure over distance from the CETRAM and/or vary by CETRAM management approach?
- Do perceptions of personal security on public transport vary according to the surrounding environment of the routes?
- What relationships exist between individual characteristics (for example, age, education), security concerns and awareness, and how do trip characteristics relate to perceived personal security?
- Do ICT technologies related to CETRAM management have a role in personal security perception among public transport users?
- Finally, do factors associated with security perceptions vary by gender?

We attempt to answer these and other questions through deployment of a smartphone-based field data collection tool in an attempt to answer the meta-question: can new ICT technologies improve urban mobility diagnostics?

3 RESEARCH DESIGN AND METHODS

To answer the research questions, we adapted our smartphone-based app originally used for mapping Dhaka's public transportation (Zegras et al., 2015). The Android app was designed in MIT App Inventor, a free, web-based service intended to make app programming easy.[7] Key enhancements to the Dhaka app included making the survey component within the smartphone app more stable (less prone to crashes), which is key to consistent surveying and matching surveys to geospatial

[7] We used App Inventor 1, the version available as of May 2013. See: appinventor.mit.edu.

and temporal information. Our app, dubbed 'Flocktracker', used App Inventor's built-in GPS call function, which attempts to geolocate the phone's position every 30 seconds. This locational functionality was not perfect, as 30-second marks were sometimes missed, either because the process timed out, was interrupted by subsequent commands,[8] and/or did not find a GPS signal. These problems hampered location accuracy during the course of the surveys. The relatively high GPS sampling frequency (30 seconds) meant that battery power with the app running non-stop sufficed for 5 to 6 hours of operation, enough to last the surveying shifts, but also a constraint.

Within the app, we designed a 12-question survey to collect interviewees' opinions regarding security (including perceptions of security at the point of the interview, importance of security in travel mode choice, perspective regarding relative security of different travel modes), technological awareness and perceptions (that is, knowledge of RFID-equipped *colectivos* and opinions regarding impacts on one's security), other relevant individual characteristics (age, gender), and trip characteristics (purpose, origin–destination, number of companions). The questionnaire was designed to be implemented on-vehicle along *colectivo* routes. Thus, speed and ease of implementation was a key specification. The app also allowed the surveyor to count the number of on-board passengers, by gender, and enabled automatic collection of location and temporal data (Figure 5.2). These data were uploaded every 30 seconds to an on-line database (as a Google Fusion table), for storage security and robustness. Not only were we able to directly associate survey and count data with spatiotemporal data, but the approach also enabled an almost real-time monitoring of the surveyors' locations (through a web-based interface, Figure 5.3), which, among other benefits, allowed surveyors' progress to be monitored and guided.

Sampling Design and Implementation

The surveys were conducted by approximately 30 university student volunteers from MIT and the National Autonomous University of Mexico (UNAM). For each of the two CETRAMs, we selected two *colectivo* routes, one which utilised the CETRAM and a second which closely follows the latter's route, but without entering the CETRAM. This design was intended to test the CETRAM versus no-CETRAM effect and possible differences in effects between the two CETRAMs, while roughly

---

8 A limitation in our use of App Inventor was that only one call function could be called at a time – for one action to occur, all others had to be stopped.

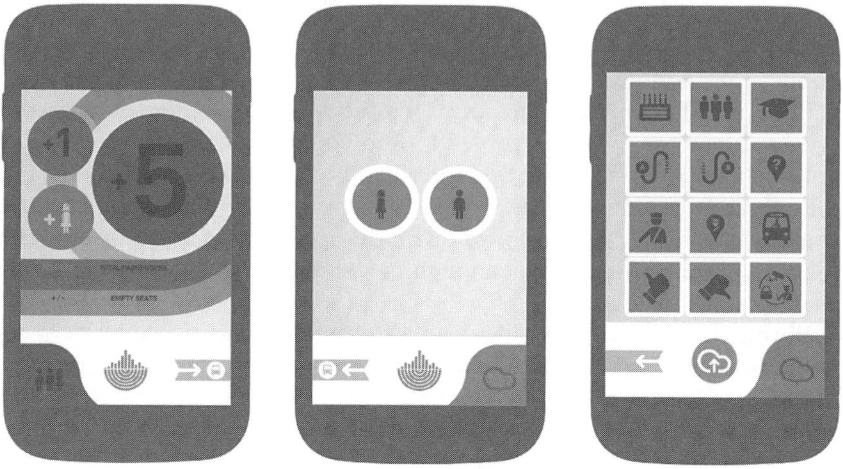

Figure 5.2 *Images of the app interfaces for (from left) passenger counting, noting gender (survey initiation), and entering individual survey questions*

Figure 5.3 *Screen capture of web-based monitoring system to locate surveyor locations while in the field*

controlling for neighbourhood-level characteristics. At El Rosario, the route was roughly 14 kilometres, while at Mexipuerto the route was roughly 16 kilometres. In the Mexipuerto case, the same company operates nearly identical routes that service the sprawling suburban neighbourhood north of the CETRAM. Some of the vehicles enter the station while others simply pick up users along the side of the broad thoroughfare adjacent to the station. Thus, the routes are essentially the same along the primary portion of the route, and all tend to meander according to demand, time of day, and reports from other operators. Figure 5.4 depicts the routes examined for both CETRAMs.

The volunteers conducted the surveys in groups of two (for personal security reasons), during one of two shifts during the day: an off-peak period from 12:00 pm to 3 pm and an evening peak period from 4:00 pm to 7:00 pm. The shifts were intended to keep volunteers from riding at night, especially once it was dark. Surveying occurred over a one-week period in mid-July, 2013. Volunteers wore uniformed shirts for identification purposes and were instructed to attempt to interview everybody on-board the vehicle. The survey required between one and three minutes to implement per observation.

Data Collection Results

The survey effort resulted in roughly 650 volunteer hours, an average of 90 hours of surveying per day, producing 10977 cleaned data points along the four routes, with 1528 interviews performed in total. Of these interviews, 1018 were performed on-board *colectivos*, the observations of interest in our analysis. The spatial and temporal data revealed little variation in travel speeds, with the average typically above 20 km per hour. Additional data collected to complement the app-based survey and vehicle condition data were observation-based, urban typology characterisation of the route surroundings. Admittedly crude, eight characterisations were operationalised, representing roadway and physical form (street types, pedestrian-orientation).

Table 5.2 presents the variables used in our analysis and their descriptive statistics. Given the importance of gender in public transportation security, we present the statistics separately for females and males. The data reveal the majority of observed trips were by *combi*, split roughly equally among the two CETRAM areas. Between one-fifth and one-quarter of the trips were on vehicles that passed through CETRAMs. The majority of the individuals were between 25 and 64 years old, with only slight variation by gender, and the majority were traveling alone, although a larger share of men than women were unaccompanied. Female travellers were more likely to be on shopping trips and men on work trips.

Expectedly, women had lower reported levels of security on their travel mode, in general, but only modestly lower levels of reported security at the point of the interview. All respondents overwhelmingly rated security as important in mode choice, again with slightly more women placing high importance on this attribute; both men and women ranked *peseros* as the

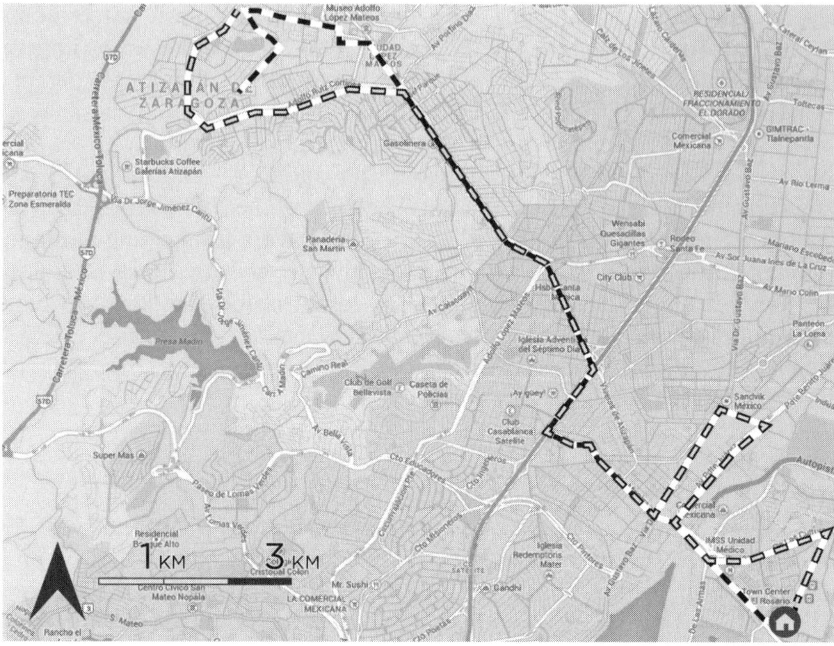

Notes: In the El Rosario map, dashed portions indicate divergent paths for route operations inside and outside the CETRAM. The solid portion represents that common to both routes. In the Mexipuerto map, the various pathways represent route deviations.

Figure 5.4 Map of surveyed colectivo routes emanating from El Rosario (above) and Mexipuerto (right)

least safe public transport mode and the Metro as the safest. Nearly two-thirds of all respondents were unfamiliar with the RFID tag system on *colectivos*; among those who knew about it, most were neutral regarding its impacts on security. More than half of respondents in the Mexipuerto service area were familiar with the RFID-equipped vehicles (56 per cent of females and 60 per cent of males). Oddly, fewer *colectivo* riders on RFID vehicles (that is, using the Mexipuerto CETRAM) knew about the RFID than those on non-RFID vehicles.

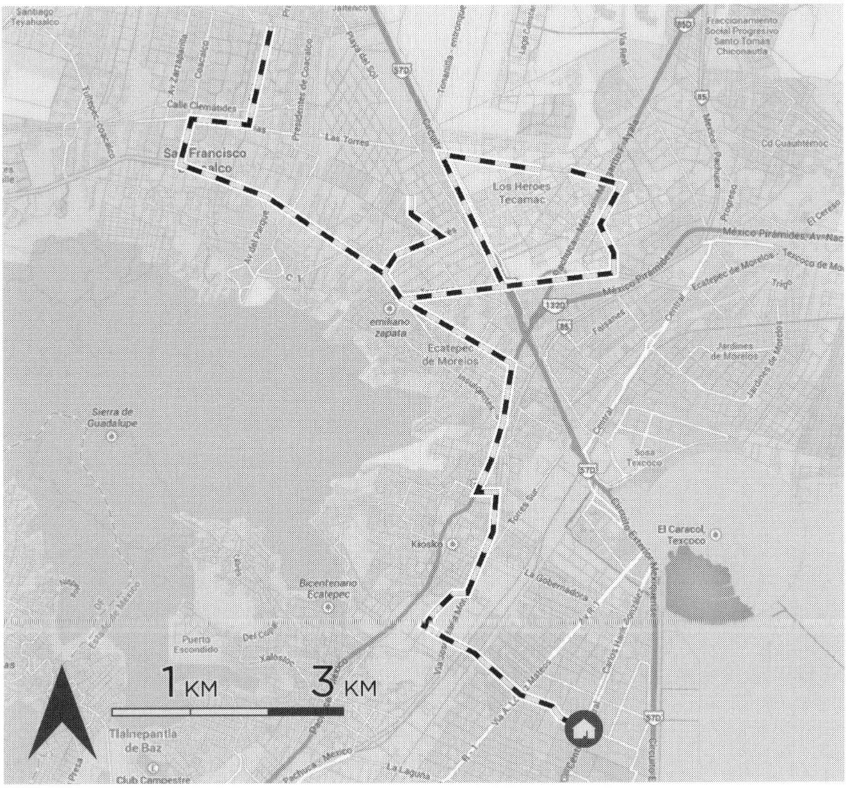

Figure 5.4 (continued)

4 MULTIVARIATE ANALYSIS OF SECURITY PERCEPTIONS

We now attempt to discern the contributing factors to on-board public security perceptions among the *colectivo* passengers in the two study areas, examining the role of CETRAM-based routes, individual characteristics, and trip-based characteristics. As our outcome, respondents' ranking of their public security level, is ordered, we specify and estimate an ordered probit model, clustering the standard errors to account for riders in the same vehicle. Table 5.3 presents the results of the best model estimated.

Table 5.2 Variables and descriptive statistics (percentage respondents or means)

Variable	Description	All	Female	Male
Combi	Trip made on *combi* (van)	0.68	0.73	0.64
Pesero	Trip made on *pesero* (minibus)	0.32	0.27	0.36
Mexipuerto	Trip in Mexipuerto study area	0.47	0.49	0.45
El Rosario	Trip in El Rosario study area	0.53	0.51	0.55
CETRAM Route	Route passes through CETRAM	0.45	0.47	0.42
Mex_Route	Route through CETRAM Mexipuerto	0.23	0.25	0.21
ER_Route	Route through CETRAM El Rosario	0.22	0.22	0.21
CETRAM Distance	Distance from CETRAM (Euclidean km) at time of interview	5.21 (5.2; 0.04; 28)	5.16 (5.1; 0.04; 28)	5.27 (5.36; 0.06; 28)
Mex_Dist	Distance from Mexipuerto (Euclidean km) at time of interview	3.47 (5.66; 0; 28)	3.46 (5.52; 0; 28)	3.47 (5.82; 0; 28)
ER_Distance	Distance from El Rosario (Euclidean km) at time of interview	1.75 (2.7; 0; 18.5)	1.70 (2.8; 0; 18.5)	1.80 (2.72; 0; 9.39)
Mex_Route_Dist	Distance from Mexipuerto (Euclidean km), route through CETRAM	1.57 (4.1; 0; 28)	1.78 (4.4; 0; 28)	1.35 (3.8; 0; 28)
ER_Route_Dist	Distance from El Rosario (Euclidean km), route through CETRAM	0.76 (2.16; 0; 18.5)	0.77 (2.22; 0; 18.5)	0.75 (2.1; 0; 9)
Peak	Trip during peak	0.5	0.51	0.49
Age_0_17	Subject is 0–17 years old	0.08	0.08	0.08
Age_18_24	Subject is 18–24 years old	0.25	0.22	0.29
Age_25_44	Subject is 25–44 years old	0.40	0.39	0.41
Age_45_64	Subject is 45–64 years old	0.19	0.20	0.18
Age_65	Subject over 65	0.04	0.04	0.03
Companions0	Subject traveling alone	0.61	0.53	0.70
Companions1–2	Subject traveling with 1–2 companions	0.27	0.31	0.24

Companions3–4	Subject traveling with 3–4 companions	0/06	0.09	0.03
Companions5+	Subject traveling with 5+ companions	0.01	0.02	0.00
Ed_Primary	Subject has primary school education	0.07	0.09	0.05
Ed_Secondary	Subject has secondary school education	0.22	0.23	0.21
Ed_HighSchool	Subject has high school education	0.41	0.35	0.47
Ed_Univ	Subject has university degree	0.23	0.25	0.22
Ed_Grad	Subject has post-graduate degree	0.01	0.01	0.01
Recreation	Subject on recreation trip	0.15	0.17	0.14
Shopping	Subject on shopping trip	0.12	0.14	0.09
School	Subject on school trip	0.07	0.08	0.07
Work	Subject on work trip	0.41	0.33	0.50
Other	Subject on other trip	0.20	0.22	0.17
Mode Security 1	Perception of security on mode (low)	0.23	0.27	0.18
Mode Security 2	Perception of security on mode	0.11	0.11	0.11
Mode Security 3	Perception of security on mode	0.36	0.33	0.39
Mode Security 4	Perception of security on mode	0.19	0.18	0.21
Mode Security 5	Perception of security on mode (high)	0.08	0.07	0.09
Point Security 1	Perception of security at current loc. (low)	0.22	0.24	0.21
Point Security 2	Perception of security at current location	0.12	0.13	0.11
Point Security 3	Perception of security at current location	0.30	0.29	0.32
Point Security 4	Perception of security at current location	0.21	0.20	0.23
Point Security 5	Perception of security, current loc. (high)	0.11	0.10	0.11
Security Impt 1	Importance, security in mode ch. (low)	0.01	0.01	0.02
Security Impt 2	Importance of security in mode choice	0.01	0.00	0.02
Security Impt 3	Importance of security in mode choice	0.05	0.03	0.06
Security Impt 4	Importance of security in mode choice	0.08	0.06	0.10
Security Impt 5	Importance, security in mode ch. (high)	0.82	0.85	0.78

Table 5.2 (continued)

Variable	Description	All	Female	Male
PeseroSafest	Subject ranks *pesero* as safest mode	0.06	0.06	0.05
TaxiSafest	Subject ranks taxi as safest mode	0.17	0.17	0.16
LRTSafest	Subject ranks LRT as safest mode	0.06	0.04	0.07
BRTSafest	Subject ranks BRT as safest mode	0.17	0.17	0.18
MetroSafest	Subject ranks metro as safest mode	0.49	0.48	0.50
PerosLeastSafe	Subject ranks *pesero* as least safe mode	0.67	0.63	0.71
TaxiLeastSafe	Subject ranks taxi as least safe mode	0.15	0.16	0.14
LRTLeastSafe	Subject ranks LRT as least safe mode	0.03	0.03	0.03
BRTLeastSafe	Subject ranks BRT as least safe mode	0.04	0.03	0.04
MetroLeastSafe	Subject ranks metro as least safe mode	0.06	0.08	0.04
RFID_No	Unaware of RFID-devices on vehicles	0.65	0.66	0.65
RFID 1	RFID has positive security impact (low)	0.03	0.02	0.04
RFID 2	RFID has positive security impact	0.01	0.01	0.01
RFID 3	RFID has positive security impact	0.15	0.13	0.16
RFID 4	RFID has positive security impact	0.06	0.07	0.05
RFID 5	RFID has positive security impact (high)	0.06	0.07	0.05
CETRAM	Immediate CETRAM surroundings	0.11	0.12	0.10
Res_Streets	Residential streets	0.30	0.32	0.27
Suburb_Comm	Large format suburban commercial	0.30	0.02	0.04
Ped_Ave	Pedestrian-friendly avenue	0.02	0.02	0.02
Auto_Ave	Auto-oriented avenue	0.15	0.13	0.16
Express	Expressway	0.22	0.21	0.24
Industry	Industrial zone	0.03	0.03	0.02
Ped_Street	Pedestrian-friendly secondary streets	0.08	0.09	0.07

Notes: n = 1018 (526 females, 492 males). For continuous variables values in parentheses, respectively: (standard deviation, minimum, maximum).

Table 5.3 Ordered probit model of colectivo riders' security perceptions

Variable	Description	Beta	S.E.	Z-Score	p-value
Survey Location: CETRAM					
Mexipuerto	Mexipuerto Study Area	−0.58	0.12	−5.00	0.00
Survey Location: Urban Typological Context					
CETRAM	CETRAM proximity	0.67	0.14	4.86	0.00
Ped_Ave	Pedestrian-Friendly Ave.	1.12	0.49	2.27	0.02
F_Ped_Ave	Female on Pedestrian-Friendly Ave.	−1.03	0.59	−1.75	0.08
Ped_Street	Pedestrian Street	0.37	0.14	2.64	0.01
Res_Streets	Residential Street	0.36	0.10	3.45	0.00
Suburb_Comm	Suburban Commercial	0.89	0.17	5.31	0.00
Industry	Industrial Zone	0.35	0.21	1.67	0.09
Express	Expressway	0.36	0.11	3.34	0.00
	Auto-oriented Ave, Ped-oriented Secondary – Reference				
Trip Characteristics					
F_companions5	Female with 5+ Companions	−0.56	0.26	−2.16	0.03
Combi	Trip on Combi (van)	0.48	0.11	4.25	0.00
F_School	Female School Trip	0.57	0.15	3.69	0.00
Trip-Maker Characteristics					
Female	Female	−1.11	0.22	−5.02	0.00
Age_0_17	Age 0–17 years old	0.22	0.12	1.80	0.07
Ed_PostGrad	Postgraduate Education	−0.34	0.20	−1.71	0.09
RFID_4	RFID has positive Security Impact (4 of 5)	0.60	0.17	3.43	0.00
RFID_No	Unaware of RFID devices	0.30	0.14	2.14	0.03
F_RFID_5	For Female RFI has positive security impact (5 of 5)	0.87	0.21	4.06	0.00
F_RFID_No	For Female, Unaware of RFID	0.40	0.17	2.41	0.02
Impt_Security_4	Importance of security in mode choice (4 of 5)	0.97	0.17	5.53	0.00
Impt_Security_5	Importance of security in mode choice (5 of 5)	0.42	0.18	2.37	0.02
F_Impt_Security_5	For Female, Importance of security in mode choice (5 of 5)	0.66	0.20	3.35	0.00

N=1018; pseudo-R^2=0.08; log pseudo-likelihood = −1531.87

CETRAM- and Urban Context-related Relationships

The results suggest that, consistent perhaps with the overall lower income of the service area, users of the Mexipuerto service area consider it less secure than do users of the El Rosario service area, irrespective of gender. The urban typology relationships are evaluated relative to auto-oriented avenues, and pedestrian-oriented secondary streets. Relative to those, the redeveloped CETRAMs' surroundings have a positive relationship with *colectivo* users' security perceptions, as do all of the other urban typological contexts. The exception to this is pedestrian avenues for women, which essentially is no different from the reference. We cannot come to any real conclusions from these results except to say that context does apparently matter with respect to public transport users' security perceptions – an area worth additional study.

Trip- and Individual-related Characteristics

For female passengers, traveling with five or more companions is associated with feeling less secure ($p = 0.03$). At least two explanations may exist for this result: females inclined towards feeling less secure may travel in larger groups; alternatively, females feeling less secure may, in fact, misrepresent their number of traveling companions as a safety precaution. Relative to work, shopping, recreational or other trips, being a female on a school trip is associated with a higher sense of security; no trip-related effects are detected specifically for men. Riding on *combis* is associated with reporting higher levels of security relative to riding on larger *peseros*. Conversations with some individuals riding the *combis* suggested that the awkward nature of entering the vehicle (sliding door had to be opened from the inside) versus the ease of access of a *pesero* and the lower capacity and relative crowding of the smaller vehicles made them harder to board and alight from and more 'intimate' for those on-board. These factors might combine to make it harder for assaults to take place on *combis*, explaining the higher perceived levels of security. Security perceptions do not vary by peak/off-peak periods.

Controlling for other individual characteristics, being a woman is associated with reporting lower personal security levels. Only the youngest age group had a statistically significant relationship ($p = 0.07$) with perceived security; this positive relationship might reflect an 'invincibility of youth' belief. For both males and females, education levels had the same effect: having a post-graduate education (albeit representing a small share of individuals in the sample) is associated with lower security perceptions ($p = 0.09$). Those who ranked security as highly important (4 or 5) in the mode choice decisions actually had a higher perceived security, perhaps reflecting travel choices that fulfilled these security predilections; the strength of this relationship is even stronger among women.

Finally, examining the Mexipuerto CETRAM-associated RFID-equipped vehicles and whether knowledge of this ICT-enabled monitoring and organisation approach had any association with *colectivo* passengers' perceived security, the evidence suggests that 'ignorance is bliss', as unawareness of the RFID devices was associated with a higher sense of security, an effect compounded for women. Being aware of the system and feeling modestly positive towards its impact on security are associated with higher levels of on-board security; again, effects strengthened for women.

Limitations and Future Research Possibilities

The findings and interpretations above regarding the CETRAMs and *colectivo* users' security perceptions should be viewed as preliminary and indicative, at best. Our fundamental objective was to demonstrate the potential value of new ICT technologies for enabling quick, high resolution urban mobility diagnostics. The approach represents an early trial, intended to chart a path for more rigorous applications. The technology proved reasonably reliable, but the findings may be overly error-prone. Survey implementation depended upon young university students with limited training, which may have impacted the sampling procedure in unknown ways. The survey instrument, designed to be quickly and easily deployed on often crowded vehicles, may be unreliable (for example, inclined towards user error) as may be the indicators themselves (for example, varying interviewee interpretations of the term 'security'). The validity of the dependent variable was also untested. Numerous potentially relevant socioeconomic characteristics, most notably, income, were excluded. While facilitating speed of implementation without discouraging respondents from participating, these omissions are likely to influence the results and certainly hamper generalisability. Similarly, we covered only a single week, during the summer break, and limited travel periods: afternoon off-peak and evening peak.

The research design itself could be improved, as we only chose four routes; the 'control' route may not have been a reasonable comparator and the companies operating each may not be more broadly representative of *colectivo* services in Mexico City. Finally, as surveyed security perceptions apparently varied widely across the extent of the corridors (Figure 5.5), our analysis would be improved with additional details characterising the physical, socioeconomic, demographic and actual security (for example, crime levels) conditions along the extent of the corridors studied.

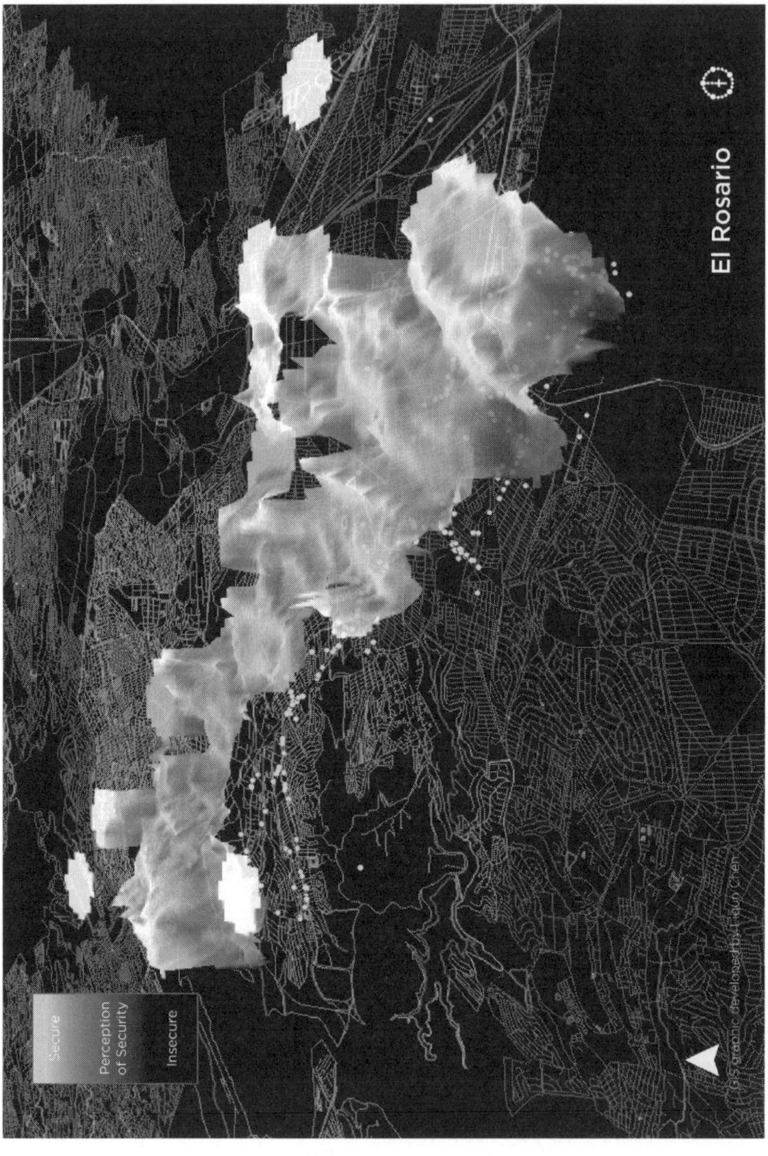
Secure
Perception of Security
Insecure
El Rosario

Figure 5.5 *Virtual 'topography' of user-reported personal security perceptions on the routes*

Although in theory the technology allows precise spatiotemporal matching of survey responses with vehicle conditions (for example, crowding), performance (for example, speeds), and context, our technological trial faced practical challenges regarding these ends. Vehicle crowding may affect security perceptions by increasing user anxiety and stress (House of Commons, 2003), but we did not compile complete crowding metrics in the course of the fieldwork. Technological limitations (that is, geolocating difficulties which 'bounced' some upload points) also prevented us from generating precise speed measures along the full extent of the corridors. These problems related to the still primitive nature of the app we deployed, which could only call one function at a time and thus faced challenges in allocating time to geolocating, carrying out surveys, or uploading data. This stands in contrast to the TransitWand implementation available at the time, which did not have on-board surveying capabilities and required manual data uploads. The manual uploads allows the app to work in areas with weak network service and allows for fewer conflicts with other actions on the device (for example, geolocating, counting), an advantage over the Flocktracker version we used. The downsides to the manual upload approach include the risk of losing data if a phone is damaged or stolen before an upload and the inability to monitor the field data collection remotely in real-time for validation and safety concerns (as in Figure 5.3).

The need for some of these tradeoffs has been reduced with the latest version of Flocktracker, which we piloted in Mexico City in January 2014. Moving beyond the AppInventor 1 built prototype, we developed a natively coded app, with more user-friendly features (including colours better suited to working in glare): an index page that enables full navigation on the app; a counter which does not impede simultaneous surveying; a statistics page to keep track of progress, location, trip attributes and so on; a customisable survey composition feature which allows research projects to easily create their own survey instruments/questions; and the ability to run persistent background processes, which among other benefits allows the Android device to sleep during inactive periods, which is an important battery saver (initial pilots reveal nearly double the battery life over our earlier version). The tool is now ready to develop richer portraits of safety perceptions, consistent with the 'whole journey approach' (Smith, 2008), as well as to aid a broad range of field-based inquiries in transportation or other sectors.

5 CONCLUSIONS

The smartphone, epitomising the ongoing global ICT transformation, promises to change the way we understand, plan, operate and use urban mobility systems. We examined the smartphone's potential as a mobility diagnostics tool, using it to measure public transport users' personal security perceptions and how they might be related to infrastructure improvements, specifically Mexico City's intermodal stations, or CETRAMs. Efforts to improve CETRAMs via real estate-oriented PPPs are consistent with relatively recent moves towards increased 'privatisation' of public safety in the city, including through the use of advanced technology. Deploying a team of volunteers armed with a purpose-built smartphone app, we collected data from over 1000 users of *colectivos* whose routes passed through or proximate to two revamped CETRAMs in the metropolis' northeast and northwest.

We find little relationship between the CETRAMs and sense of security among users of *colectivos*; passengers on vehicles in proximity to CETRAMs report higher personal security levels, but those positive context relationships are similar to those associated with a range of other contexts where the on-board surveys were conducted. We do find a range of contextual settings relate positively to security perceptions, but more research will be required to understand why. Those believing strongly in the safety benefits of RFID-equipped vehicles report higher security levels on board (whether the vehicle has RFID or not); this modest ICT technology is associated with even higher levels of security among women, who view it favourably. Younger passengers report higher levels of personal security, while the well-educated do not. Unsurprisingly, gender matters, with female respondents in general reporting lower levels of security, and female-specific effects detected for trip, context and security perception-related variables.

While showing the viability of the smartphone as an urban diagnostic tool, the research faces numerous limitations, which suggests the above results should be viewed as preliminary. The survey was limited in temporal scope, implemented by student volunteers, and collected the bare minimum of information from *colectivo* users. The validity and reliability of measures for the dependent variable can be questioned, and additional contextual information (speeds, physical and social surroundings) should be incorporated into future analyses. The newer version of the app we deployed for this data collection represents a major advancement, which we aim soon to make widely available and envision it ultimately being deployed by anybody with interests in better understanding urban phenomena. In our continuously urbanising world, it is hoped that increasingly powerful ICT technologies like the smartphone can be used to better diagnose and develop urban mobility systems.

ACKNOWLEDGEMENTS

This Chapter is based upon research produced through field work supported by a grant from MISTI Mexico and in a partnership with Urban Travel Logistics (UTL). Additional research support came from the Singapore National Research Foundation under the SMART Future Urban Mobility Research Group. We thank Florencia Serrania and Isaac Lozada of UTL for all the support in carrying out the field work. Special thanks also to Gabriel Hernandez of UNAM, Griselda Gomez at MIT-Mexico, MIT graduate students Liqun Chen, M. Bin Jung, and Clara Suh, for their collaboration and contributions and the numerous volunteers who helped carry out the field work.

REFERENCES

Berg, M., P. Pokharel, A. Sweetland and V. Modi (2013), 'Cellular Citizenship', *Harvard International Review*, **34** (3), 28–32.

Cottrill, C., F. Pereira, F. Zhao, I. Dias, H.B. Lim, M. Ben-Akiva and C. Zegras (2013), 'Future Mobility Survey: Experience in Developing a Smart-Phone-Based Travel Survey in Singapore', *Transportation Research Record*, **2354**, 59–67.

Davis, D. (2007), 'Urban Violence, Quality of Life, and the Future of Latin American Cities: The Dismal Record So Far and the Search for New Analytical Frameworks to Sustain the Bias Towards Hope', in A. Garland, M. Massoumi and B. Ruble (eds), *Global Urban Poverty: Setting the Agenda*, Washington, DC: Woodrow Wilson International Center for Scholars, pp. 57–87.

Davis, D. (2013), 'Zero-Tolerance Policing, Stealth Real Estate Development, and the Transformation of Public Space: Evidence from Mexico City', *Latin American Perspectives*, **20** (2), 53–76.

Eros, E., S. Mehndiratta, C. Zegras, K. Webb and M.C. Ochoa (2014), 'Applying the General Transit Feed Specification (GTFS) in the global south: Experiences in Mexico City and beyond', *Transportation Research Record*, **2442**, 44–52.

Gobierno del Distrito Federal (GDF) (2011), 'Evaluación del diseño e instrumentación de la política de transporte público colectivo de pasajeros en el Distrito Federal', accessed 18 January 2014 at: http://www.evalua.df.gob.mx/files/recomendaciones/evaluaciones_finales/ev_transp.pdf.

González, M., C. Hidalgo and A-L. Barabási (2008), 'Understanding Individual Human Mobility Patterns', *Nature*, **453**, 779–82.

House of Commons (2003), *Overcrowding on Public Transport. Seventh Report of Session 2002–03*. Volume I, 17 September, London: The Stationery Office.

Hsu, H-P. (2011), 'How Does Fear of Sexual Harassment on Transit Affect Women's Use of Transit?' *Women's Issues in Transportation: Transportation Research Board Conference Proceedings*, **46** (2), 85–94.
INEGI (2000), *Distrito Federal, XII Censo General de Población y Vivienda 2000*, TabuladosBásicos, Mexico City: National Statistics Institute.
INEGI (2010), *Censo de Población y Vivienda 2010*, Mexico City: National Statistics Institute.
Jariyasunant, J., M. Abou-Zeid, A. Carrel, V. Ekambaram, D. Gaker, R. Sengupta and J. Walker (2014), 'Quantified Traveler: Travel Feedback Meets the Cloud to Change Behavior', *Journal of Intelligent Transportation Systems* (forthcoming).
Lynch, G. and S. Atkins (1988), 'The Influence of Personal Security Fears on Women's Travel Patterns', *Transportation*, **15**, 257–77.
Newton, A. (2004), 'Crime on Public Transport: "Static" and "Non-Static" (Moving) Crime Events', *Western Criminology Review*, **5** (3), 25–42.
SETRAVI (2010), *Programa Integral de Transporte y Vialidad 2007–2012*, Mexico City: La Secretaría de Transportes y Vialidad del Gobierno del Distrito Federal.
Smith, M. (2008), 'Addressing the Security Needs of Women Passengers on Public Transport', *Security Journal*, **21**, 117–33.
Sutton, R. and S. Farrall (2005), 'Gender, Socially Desirable Responding and the Fear of Crime: Are Women really more Anxious about Crime?', *British Journal of Criminology*, **45** (2), 212–24.
Yavuz, N. and E. Welch (2010), 'Addressing Fear of Crime in Public Space: Gender Differences in Reaction to Safety Measures in Train Transit', *Urban Studies*, **47** (12), 2491–515.
Zegras, P.C., E. Eros, K. Butts, E. Resor, S. Kennedy, A. Ching and M. Mamun (2015), 'Tracing a Path to Knowledge? Indicative User Impacts of Introducing a Public Transport Map in Dhaka, Bangladesh', *Cambridge Journal of Regions, Economy and Society*, **8** (1), 113–129.

6. Enhancing transport through open geospatial data and crowdsourcing*

Qiyang Xu and Elizabeth Dodds

1 INTRODUCTION

While developed countries have traditionally led the transformation in the era of rapid technology development, medium-income and low-income countries have joined the burgeoning markets of ICT and are expected to outpace high-income countries in adopting these technologies. The significant role of ICT is also recognized in a way that is relevant to tangible aspects of life, such as in emergency response and transport planning for commuting purposes. As a crucial component of good governance, maintenance of social stability and a path to balanced economic growth, transport and ICT services serve as an important intermediate component linking development resources to those in need. During the Organisation for Economic Co-operation and Development's POVNET workshop on infrastructure and poverty reduction in 2004, the International Development Center of Japan (2004) reported on the close relationship between transport and ICT in increasing labour and job demand during the construction period and in promoting productivity after infrastructure is in place. Furthermore, their findings indicate that combining ICT with transport initiatives could make resources, career opportunities and markets more accessible to the poor.

As urbanization and transport infrastructure development are still in the early phases of expansion in developing countries, the challenges facing the transport sector differ significantly from those of developed countries. While transport initiatives in developed countries tend to focus on increasing efficiency of fuel consumption and preventing pollution from fossil fuels (Geenhuizen, 2009), the huge market potential of ICT-enabled transport and commuting planning in the setting of developing

* The opinions and the information presented in this chapter are the sole responsibility of the authors, and in no way reflect the policies of the World Bank. Any remaining errors and omissions are the sole responsibility of the chapter authors.

countries means that these countries are more concerned with the application of ICTs to:

- mitigate massive transport issues caused by the rapid pace of urbanization;
- expand transport networks to provide safe and reliable access to urban and less developed areas of the country, with the main objective of sustaining economic development in the long term.

In the past decade, ICT has contributed significantly to crafting efficient traffic management systems, reducing traffic congestion and enhancing policy making and governance in the transport sector in developed countries. In particular, these countries have already benefited from the growing prominence of Global Positioning Systems (GPS) and open transportation datasets published under a universal standard – Global Transit Feed Specifications (GTFS) – created by Google. This chapter focuses specifically on how open geospatial data and crowdsourcing can support ICT-enabled innovations for transport in a global context with emphasis on developing countries. In section 2, we discuss open transport data and crowdsourcing, two major innovative ICT applications. Section 3 discusses the challenges of adopting open data and crowdsourcing mechanisms in generating and sharing transport data, as well as the digital divide in developing countries. Finally, section 4 provides recommendations for policymakers and practitioners interested in this field in the future.

2 OPEN DATA AND CROWDSOURCING FOR EFFICIENT TRANSPORT PLANNING

2.1 Leveraging Open Transport Data Usage under a Common Standard

A fundamental objective of creating an efficient transport system is to offer reliable traffic information to passengers. Concerns over whether a lack of data transparency impedes software innovation and social accountability, combined with greater demand for public access to data traditionally hosted by local or national authorities have led to a new direction in data literacy. The term 'open data' was introduced in 1995 by an American scientific agency document to refer to the sharing of geophysical and environmental data freely as common goods. It embraces the idea of enabling public access to, free use and reuse of datasets released by multiple stakeholders including government agencies, research centres or other sources without restricting their use by expensive licences or other bureaucratic controls (Auer et al., 2007). This argument does not advocate that all data should become freely available to anyone for any purpose, but rather that certain existing data could be utilized to benefit the wider public, taking into account certain sensitivities.

During the first decade of the twenty-first century, governments worldwide have been promoting open data initiatives at various levels under eight principles of open government data: 'complete, primary, timely, accessible, machine-readable, non-discriminatory, non-proprietary and license-free' (Lessig, 2007). In 2010, the UK government and Rennes, a medium sized city in France, launched Open Data Initiatives, followed by Berlin in 2011. Whilst the Organisation for Economic Co-operation and Development countries played a leading role in Open Data Initiatives, developing countries such as Kenya, Uruguay and Moldova launched open data campaigns as well.

One important category of open data is transport data and the associated geospatial information. A 2009 study on the contribution of spatial information to New Zealand's economy found that open spatial data contributed at least 0.6 per cent to national GDP, equivalent to US$1.2 billion. The study prepared for Land Information New Zealand, the Department of Conservation and the Ministry for Economic Development found that in the transport sector, opening up geospatial information led to a decrease in fuel costs, accidents and emissions, and improved travel behaviour (SKM, Ecological Associates and ACIL Tasman, 2009). The 2010 Open Government Benchmark Study covering 1000 adults in the United States, 300 self-identified government employees and 50 civic application developers, revealed that transport data, including parking, transit and traffic, was rated as 'very important' in high value data categories by 34 per cent of citizens, 67 per cent of governments and 67 per cent of developers (Socrata et al., 2011).

The majority of academic research conducted during the late 1980s and early 1990s was reported in non-digitized formats and the same was true for transport-related data, but this has changed significantly with the widespread use of ICT in both developed and developing countries. Yet one of the major challenges for data sharing is to interpret the meaning of datasets precisely, from underlying conceptual models to structure, and to make them transferable along with the content of the data itself (Mandel and Ruffert, 2000), regardless of the various approaches, sources or transport models used to collect or generate such data.

One of the earliest efforts to create a universal standard for transport data to break down the barriers in data communication and sharing was made by Mandel and Ruffert in 1999–2000 when they developed the Generalised Transport-data Format (GTF) to mitigate the difficulty in communicating between different transport datasets, based on a thorough survey of conceptual models and formats conducted by Nielsen et al. (1998). As a data model, GTF pieces together transport information and turns it into building blocks with common standards for a holistic picture of transport entities, relationships and attributes. The General Transit Feed Specification (GTFS), a transport dataset of standards for web-based platforms or mobile applications later developed by Google, is a similar data standard covering basic transport information, including agency, timetable, stops, routes and travel time (Google, 2012). It is known as a simple yet universal standard for structuring datasets for basic transit system services, which has been adopted by over three hundred transport agencies worldwide, demonstrating the significance of common standards in this field. A good example illustrating the rising movement of open transport data and usage of GTFS is the launch of City-Go-Round. This US-centred website was developed by software engineers and open data advocates in an effort to encourage agencies to release transport datasets, especially transit schedule data, under GTFS. Meanwhile it promotes reuse of GTFS-standard transport datasets provided by some of the 740 domestic agencies listed on the website in an effort to continue to improve public transport conditions.

The creation of GTFS not only allows public access to transit schedules and better trip planning through mobile applications and data visualization, but also enables analysis of transport flows for better assessing and planning transit networks (Antrim and Barbeau, 2013). The open transport datasets generated under GTFS are incorporated into Google Maps and, more importantly, travellers can receive push and pull text or voice messages about real-time transit information through their smartphone or other ICT devices, which can potentially reduce commuting time by making trip planning more efficient while relieving traffic congestion.

Despite these developments, however, some fundamental questions remain to be answered, such as how travellers on various transport modes can obtain and use information generated by others to ultimately improve overall resource efficiency (Halonen, 2012).

2.2 Crowdsourcing Mechanism

Crowdsourcing, as the word indicates, is an approach that draws upon citizens or a targeted population to contribute to the creation and validation of certain datasets. It is increasingly gaining prominence in mapping geospatial information where individuals interact with datasets (Brigham et al., 2013; Heipke, 2010) after people have started to realize the incomparable value of information and knowledge exchange in breaking down locations and cost barriers by establishing a free marketplace for such resources (Howe, 2006). Crowdsourcing is built on the notion that the higher the number of people contributing information, the higher the accuracy and closer to reality this information gets.

Crowdsourcing occurs around various types of information and is becoming more prominent in creating, collecting and sharing geographically relevant information (Brigham et al., 2013) contributed directly by citizens and linked to their feedback. In this sense, crowdsourced geospatial information and social networks may also be referred to as 'volunteered geographic information' (VGI) (Goodchild et al., 2007). For example, in the case of OpenStreetMap, the world's largest open source mapping platform built on publicly contributed VGIs, the number of participants contributing to creating transport networks on OpenStreetMap has rapidly expanded by over ten times in the past five years, from over 100000 participants in March 2009 (Chilton, 2009) to 1.4 million in September 2013 (Wikipedia, 2013). While OpenStreetMap contributors used to be particularly active in Europe (Haklay and Weber, 2008), the coverage of OpenStreetMap is expanding globally lately, although the actual number of users still remains unidentified.

When VGI is generated by ICTs such as SMS or GPS on vehicles, crowdsourcing can be used to contribute to publicly managed transport databases through VGI on transport issues such as rail capacity, road traffic conditions or emergencies. This contribution can offer a real-time low-cost snapshot of the transport system by increasing access to information as well as offering more regular updates. In order to reach as many users as possible, transport VGI needs to be shared publicly through open data initiatives and under a common standardized traffic data format that is not only accessible, but also interpretable and reusable locally as well as by commuters in the wider area.

One field where crowdsourcing geographic and transport network information is particularly valuable is disaster response and relief. The 2008 Haiti earthquake was a watershed event that made the country visible to the rest of the world. The blank spot on OpenStreetMap was later filled with a complete illustration of the whole country by crowdsourcing web-based mapping platforms such as Ushahidi, GeoCommons and CrisisCamp Haiti (Goodchild and Glennon, 2010; Zook et al., 2010). Meanwhile the power of social media also helped to amplify the voice of the natural disaster victims and to visualize emergency reports on interactive mapping platforms so that emergency services could be directed according to the actual needs. RescuePH in the Philippines and Sinsai platform in Japan are two other successful examples of utilizing crowdsourcing for disaster response in vulnerable communities.

Crowdsourcing in conjunction with ICTs and social media bridge the gap between highly skilled map makers and actual map users, organized professionals and groups working in the field of geography and individuals, sophisticated surveying equipment and regular ICTs such as desktop computers and smartphones (Haklay and Weber, 2008). Compared to the traditional approaches of transport data collection, task assignment and management, and later to compare the completion and promptness of datasets collected, crowdsourcing implements the 'wisdom of the crowds' (Surowiecki, 2005) by encouraging users to be information contributors as well as consumers, given that time-saving is an incentive to strengthen user behaviour. Based on the facts mentioned previously, there has been contemporary interest on employing crowdsourcing in Intelligent Transport Systems (ITS) and creating a more sustainable transport system by influencing behaviour in traffic flows (Teeuw et al., 2012; Zhao, 2000) through providing crowdsourced route networks as static information and real-time traffic conditions as reference for commuting planning.

Waze, a successful smartphone application for drivers, has gained worldwide popularity by providing real-time crowdsourced traffic information to drivers in a cost effective way. Starting as a start-up company in Israel in 2008 and built based on GPS technology, Waze keeps the fundamental functionality shared by many other transport navigation applications – guiding drivers on a turn-by-turn basis. However, what makes it different from traditional GPS devices or applications is that it is also built on a community of drivers themselves, where users receive alerts not only on dynamic transport information shared by other drivers passing by, such as traffic conditions, road maintenance and accidents, but also points of interests documented by the application such as landmarks and fuel stations offering cheapest fuel along the routes. Growing for five years, Waze has evolved into a complete basemap for 14 countries in North America, South America, Europe, Africa and Asia (Waze, 2012) with over 50 million users by 2013 (Yahoo, 2013).

Viewing the transport sector in developing countries as a corporative ecosystem instead of a stand-alone system (Miller, 2013), the crowdsourcing mechanism facilitates sustainability by maximizing cooperation instead of competition among transport resource users. Built on geospatial technology and open data, it also emphasizes the power of networks, an add-on to time and space, as a catalyst of such cooperation. With relevant transport and associated geospatial information being disclosed by authorities under Open Data Initiatives, users are now actively included in the process of information sharing and are gradually beginning to take ownership of and responsibility for such activities. This phenomenon was noted by Surewiecki (2005) as leveraging the 'wisdom of the crowds'.

3 CHALLENGES OF IMPLEMENTING ICT-ENABLED INNOVATIONS

Despite the prospective future benefits of leveraging citizen participation and ICT to augment transport data, one still has to look at the challenges and pitfalls around open data, crowdsourcing and the implementation of ICT innovations in the transport sector, particularly in a developing-country context.

3.1 Accuracy of Open Geospatial Data

When it comes to open data, various issues could emerge from different aspects, ranging from basic data-sharing policy frameworks, to data licensing and data storage. No matter what data standards are adopted, data quality is always of primary importance in order to 'provide consistent results, deliver higher quality products, improve decision making processes and reduce operational cost' (Open Geospatial Consortium, 2008). One question relevant to many stakeholders is to what extent open data is accurate, the question that is closely linked to citizen trust on reusing the datasets for further analysis. Xia (2012) proposed 15 categories of requirements that an open dataset should meet, including accessibility, authority and integrity, among which accuracy comes first. Tauberer, in his book *Open Government Data*, suggests measuring the validity of open data by both precision and accuracy. The former represents the resolution of various formats of datasets, such as video, audio, numerical sheets and imagery, indicating the volume of information and knowledge stored, while the latter shows at what degree the data reflects the real world or truth (Tauberer, 2012). Among the few existing studies to date about the accuracy assessment of open geospatial datasets, the road networks analysis presented below attempts to evaluate the quality of open geospatial data released by governments.

Despite the fact that Open Data Initiatives represent major channels for public data sharing, traditional geospatial data warehouses still remain active. The study assesses geospatial information in public geospatial data portals and OpenStreetMap, particularly road networks, by comparing the routes with the ground conditions shown by high-resolution satellite imagery overlapped with datasets. The purpose of the study has been to better understand the accuracy of public transport infrastructure datasets in the context of developing countries.

3.2 Methodology

The study selected four countries, Haiti, Laos, Senegal and Togo, prioritized according to their disaster vulnerability ranks in Global Facility for Disaster Reduction and Recovery, GFDRR Post Disaster Needs Assessment (PDNA, https://www.gfdrr.org/node/118) and Risk Assessment (RA, https://www.gfdrr.org/node/65), provided by the World Bank's disaster relief unit, Global Facility for Disaster Reduction and Recovery. The four countries are highly-exposed disaster hotspots, earmarked with risk assessment and local capacity, yet they are developing countries with expanding transport infrastructures not only for disaster recovery and management but also economic growth. As the team started to search for data through multiple governmental or official repositories, it found out that publicly accessible transport data is quite scarce in these countries, causing them to rely on transport datasets from Digital Chart of the World available in diva-gis.org for validation and road features extracted from OpenStreetMap for reference. Digital Chart of the World, published in 2006, depicts various thematic geospatial features aggregated from National Imagery and Mapping Agency data between 1960 and the 1990s. Both datasets are key transport nodes and have been created with aggregation of pieces of data from different sources collected using different methodologies, which makes the whole task more challenging and interesting as a learning experience (see Figure 6.1).

This study assesses the quality of transport networks datasets from two aspects: completeness and precision. The completeness of publicly available datasets is assessed by a comparison between the total length of roads in US miles in Digital Chart of the World with that in OpenStreetMap (see Figure 6.2). The study selected the five districts in each country that have the most road features. With regard to precision, the open dataset is validated with the ground condition shown in Google Earth satellite imagery.

Each road feature in Digital Chart of the World sampled is assigned a precision code based on the extent to which it matches satellite imagery:

- Code 1: the route in satellite imagery aligns with the real conditions on the ground.
- Code 2: the route is a similar shape or direction to that shown in imagery apart from a certain distance adopted as a buffer zone.
- Code 3: the route is completely unaligned with that in satellite imagery.

The authors also indicate the degree of the match:

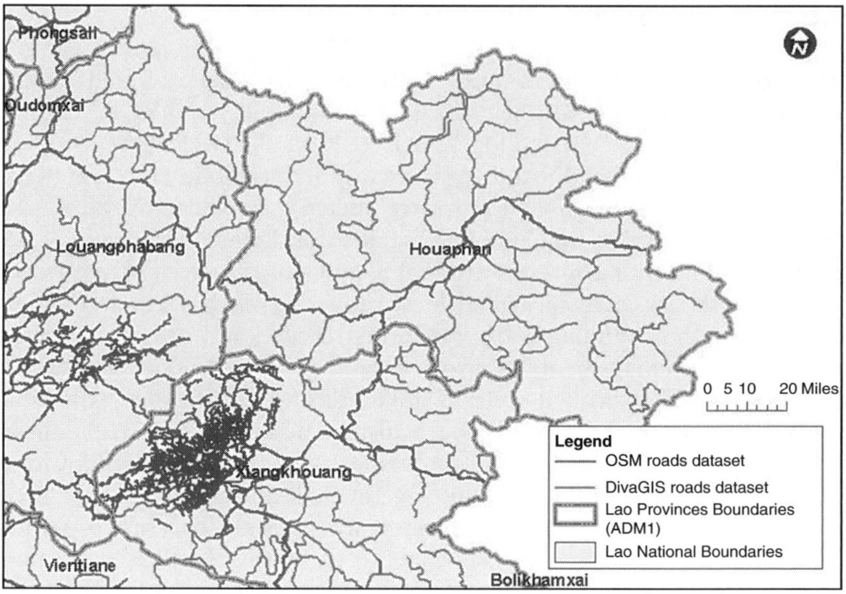

Figure 6.1 Dataset representation in Laos

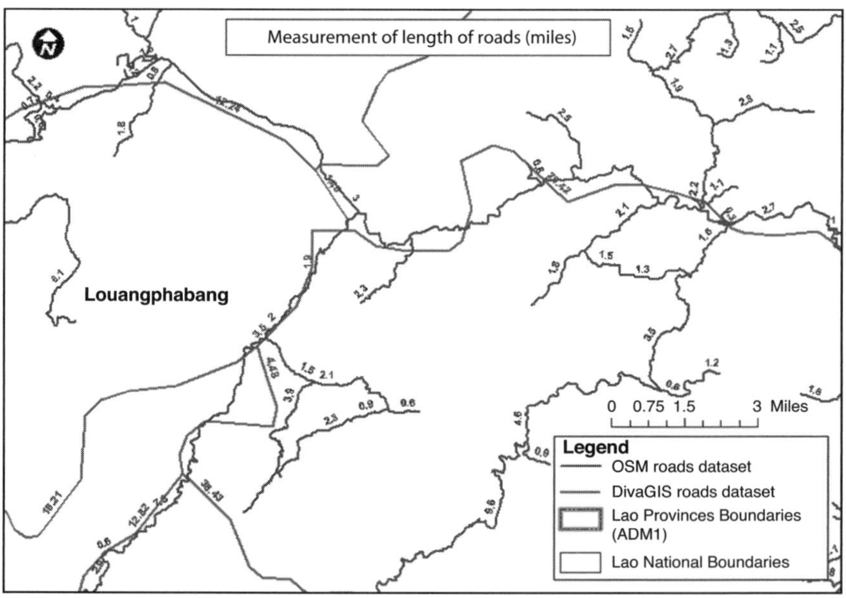

Figure 6.2 Example of road length measurement in Laos

Source: Google Earth.

Figure 6.3 *Road completely matches ground conditions*

Source: Google Earth.

Figure 6.4 *Road is close to actual path*

- Completely match: the route in satellite imagery aligns with the real conditions on the ground.
- Close to roads: the route is a similar shape or direction to that shown in imagery apart from a certain distance adopted as a buffer zone.
- Not match: the route is completely unaligned with that in satellite imagery.
- Ground condition unclear: unable to identify the transport networks due to cloud disturbance, low quality of imagery or other reasons.

Figures 6.3, 6.4, 6.5 and 6.6 illustrate the four categories mentioned above.

The study is based on two assumptions:

1. All road networks in DIVA-GIS OpenStreetMaps are traceable in satellite imagery provided.
2. Validation results of the cleaned datasets can be generalized to the whole health facility and primary school dataset, even on a broader perspective to inform geospatial information of social infrastructure in other open datasets.

Source: Google Earth.

Figure 6.5 Road passes over water

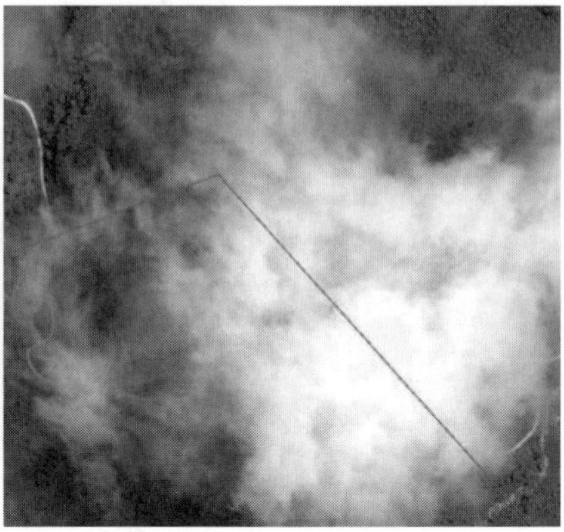

Source: Google Earth.

Figure 6.6 Ground conditions unclear

3.3 Underlying Assumptions and Limitations in the Datasets and Methodology

Unlike countries launching Open Data Initiatives and releasing governmental datasets through a single national portal, the four countries in this analysis have multiple public data portals, including ministries, statistics bureaux and government agencies. Considering that the objective of this analysis is to evaluate the official dataset that is of the best quality in the countries under study, more than one official data source can be documented; however, only the one that is more complete and accurate is used for this analysis. Therefore only open-source datasets are used in this analysis. Outdated datasets are not likely to correctly reflect the current facilities in the country. For example, the primary source for road networks, DIVA-GIS Roads Data, dates back to the early 1990s, despite the fact that it encodes roads in detailed categories, that is, primary, secondary and tertiary routes. Yet datasets from DIVA-GIS are of better quality than some of the government-provided resources.

Aside from that, the positioning accuracy of satellite imagery and the error of coordinates measured using GPS devices in the Africa region, which can reach up to six metres (Walpersdorf et al., 2007), will also potentially lead to disparities between source imagery and actual ground conditions. The vegetation in South Asia covering roads, along with time lags between Google satellite imagery, which is mostly collected after 2008, and the time when the transport network data is created, may

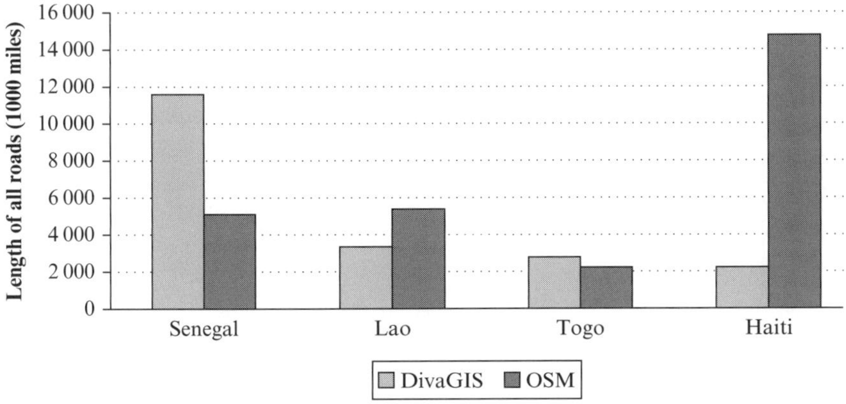

Figure 6.7 Total length of road networks in Digital Chart of the World and OpenStreetMap

contribute to incompleteness and inaccuracy. This is an important issue to consider about the use of ICT for transport, particularly in a developing-country context.

3.4 Findings

Using OpenStreetMap as a reference, the analysis shows that among the four countries mapped in Digital Chart of the World, Haiti and Senegal have the lowest and highest completeness rates respectively. While Haiti is the most mapped country by OpenStreetMap, Togo is the least among the four. With Figure 6.7 showing the total length of roads mapped for each country in Digital Chart of the World and OpenStreetMap, it has to be noted that the size of a country profoundly influences the size of the datasets. In this sense, Figure 6.8 quantifies the results proportionally and better visualizes the completeness of Digital Chart of World.

When it comes to precision, the results reveal large inconsistencies between the truth on the ground and the Digital Chart of the World dataset, whose average total length of transport networks is 19.86 per cent shorter than that in OpenStreetMap. The largest negative discrepancy is in Haiti, with a total length of roads six times less compared to OpenStreetMap data. One explanation is that in 2008 the OpenStreetMap community in Haiti created a comprehensive transport network datasets for the country to aid in post-earthquake recovery. None of the four countries have 100 per cent precision, as anticipated, and most roads in Haiti, Laos and Togo mapped by Digital Chart of the World are close to ground

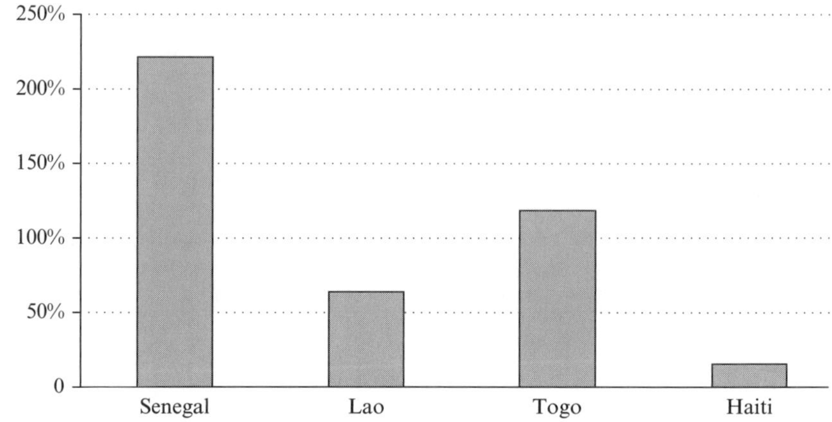

Figure 6.8 *Discrepancy in completeness between Digital Chart of the World and OpenStreetMap in the four countries (OpenStreetMap = 100%)*

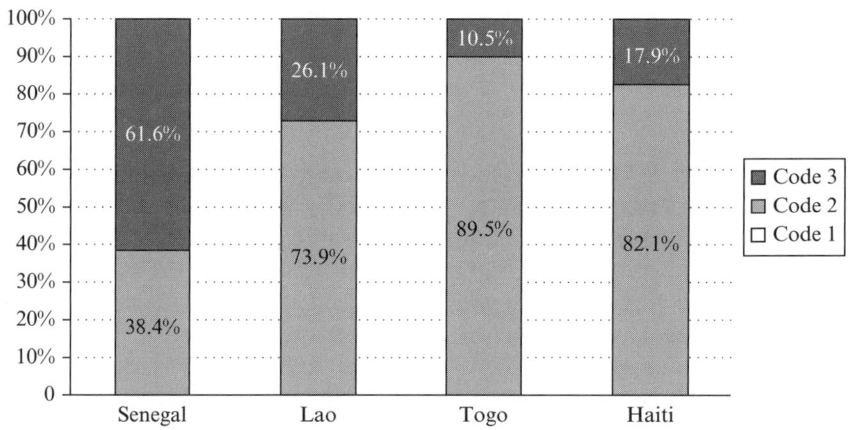

Figure 6.9 *Precision of Digital Chart of the World, compared to Google satellite imagery*

conditions indicated by imagery, but not precisely matched. The highest accuracy level is in Togo, with 89.5 per cent of the roads close to the actual ground conditions, compared to the lowest in Senegal, with an average accuracy level of 38.4 per cent (Figure 6.9).

While the study casts further doubt on the quality of open geospatial datasets, it also explores potential explanations behind it. One of the questions still unanswered is how much such studies could be generalized to other datasets while sometimes half of the datasets will have to be eliminated during data processing and database clean-up. Second, the study assesses evaluation at a national level without clearly distinguishing between urban and rural areas, while it could easily be concluded that roads sprawling in urban areas could potentially lead to a surge in accuracy. Similar to road networks, OpenStreetMap, as one of the most prominent crowdsourcing projects, gains comparatively more resources in urban areas to have transport networks mapped and the completion is considerably higher than that of rural areas in developing countries. These findings are of great importance, not only for transport infrastructure planning, but also for the use of ICT as reliable tools in such tasks to inform decision making in both developing and developed countries.

3.5 Accuracy of Crowdsourced Data

During the early days of crowdsourcing mechanisms, the importance of open geospatial data quality was gradually realized by a small group of data consumers and contributors to crowdsourced datasets in OpenStreetMap and transport navigation applications. Several reasons can impact on the reliability of crowdsourced data or VGI, for example varying user judgement about particular details of data reporting, lack of legal frameworks or consistent methodology and approaches during data collection. Given such risks it is unclear how data users could gain universal trust on crowdsourced datasets or VGI which may be available globally.

Although recent research emphasizes the unlikelihood of obtaining 100 per cent accurate geographic information (Graham, 2013), the quality of these datasets remains a crucial component in both metadata and the dataset itself. Several efforts have been made to assess data quality of the most significant crowdsourced transport route data platform, namely OpenStreetMap, while OpenStreetMap also conducted its own data assessment. The quality assessment of the OpenStreetMap dataset in France (Girres and Touya, 2010) took eight elements of spatial data quality (Guptill and Morrison 1995; Kresse and Fadaie, 2003), including geometric accuracy, attribute accuracy, completeness, logical consistency, semantic accuracy, temporal accuracy, linage and usage. Variation in quality, meta-quality and resolution were added later to the quality elements (Oort, 2006) and assessment of the positional accuracy quality (Ather, 2009; Haklay, 2010), which has continued in Indonesia (Universitas Gadjah Mada, 2013). These assessments stress the problematic heterogeneity of OpenStreetMap data due to different data sources and data collection processes despite the widespread use of OpenStreetMap.

Regardless of the accuracy issues, crowdsourcing outpaces traditional data collection approaches in terms of currency and its ability to capture ongoing changes in a more timely manner. Goodchild and Glennon (2010) pointed out that crowdsourcing would be more reliable in more densely populated geographic areas, and the quality of crowdsourced datasets could thus be improved as the information could be identified by taking consideration of the surroundings as a part of the VGI quality assurance exercise. This also echoes the fact that studies about crowdsourcing are undertaken in the context of urban areas in developed countries, while the digital revolution and emerging challenge of designing sustainable transport networks are occurring largely in the developing world.

3.6 Digital Divide: The Challenge of Digital Exclusion

While the use of ICTs in efficient transport initiatives has the potential to reduce the cost, time and distance of communicating information and to encourage broader citizen participation, the social challenges of applying ICT in the developing-country context must also be addressed (Thomopoulos and Karanasios, 2014). Low penetration rates of new technologies and significant barriers to ICT access threaten to exclude certain groups from participating in and benefiting from ICT-enabled initiatives. For this reason, it is important to consider socioeconomic, geographic and political contexts in the design and implementation of ICT-enabled transport projects to ensure they achieve, or avoid undermining, their overarching development objective.

Although the adoption of technology has rapidly increased throughout the developing world, it has also been highly uneven across and within low-income countries. The resulting disparity in access to and use of ICTs is commonly referred to as the 'digital divide'. As a result of the digital divide, one of the major concerns regarding the broad adoption of ICT in developing countries is the potential exclusion of societal groups with limited access to ICT due to the technical, economic or social barriers they face. These barriers can arise due to a lack of financial resources or human capital required to effectively interact with ICT, and often lead to a gender, socioeconomic, age and rural–urban imbalance in the use of ICT in developing countries. A common concern is that elderly, less educated and female citizens in remote areas will be excluded from participating in ICT-enabled schemes.

The digital divide is a topic of interest among development practitioners given the risk that ICT-enabled interventions may exacerbate already existing social inequities within developing countries. Particularly among poorer, rural communities, insufficient broadband access, limited mobile penetration or the relatively high cost of using ICT are likely to exclude such communities from providing input into or benefiting from an initiative that relies on more advanced technologies. Additionally, more advanced ICT such as VGI, while offering the benefit of increased efficiency and social participation, also requires greater user capabilities, in particular digital literacy, to enable citizens to interact with ICT and benefit from the information being communicated. Even in contexts with relatively widespread access to ICT, it is also important to consider the motivation of different user groups to access and use ICT as part of a development initiative. In some cases, motivation among certain groups could be low due to the relatively high cost or low perceived benefit of using the technology. For these reasons, there is the risk that ICT-enabled initiatives in developing countries will be subject to elite capture, whereby participation is dominated by already advantaged societal groups.

Particularly in the case of ICT-enabled crowdsourcing or open data initiatives in developing countries, the presence of a digital divide implies that the information collected via ICT is likely to be subject to biased input. Take for example the application of crowdsourcing to collect information on traffic patterns. In the presence of a digital divide, the accuracy of the information collected will depend on whether that population has sufficient access, resources and motivation to provide information. In the context of developing countries, there is the risk that crowdsourcing in transport initiatives will only reach a narrow spectrum of the overall affected population due to barriers in ICT access.

It is important to recognize, however, that the opportunities and challenges associated with the application of ICT to transport initiatives are highly context dependent. The term ICT encompasses a growing set of technologies with varying degrees of penetration depending on the country, region, culture and population of interest. In terms of the global growth in ICT access, Figure 6.10 demonstrates how the growth in mobile phone subscriptions has greatly outpaced the use of other ICT devices. This reflects the case for many developing countries, where mobile phone penetration is increasing rapidly, while Internet access remains limited in many areas and among many groups.

4 MOVING FORWARD

In sum, open data and crowdsourcing are accelerating the formation of a more transparent, interactive and real-time transport information

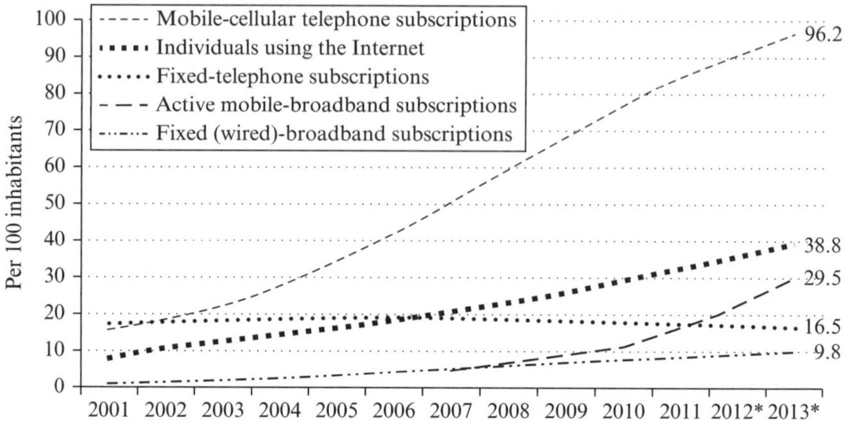

Note: * Estimate.

Source: ITU Telecommunications/ICT Indicators database.

Figure 6.10 Global ICT developments, 2001–13

network, connecting commuters based on geolocation information, creating standards for transport data and facilitating greater participation in information sharing. However, significant challenges remain, including how to ensure and enhance user interaction and reliability, as well as data accuracy. Table 6.1 provides an overview of the potential roles of different stakeholders to promote the usage and improve the quality of open data in order to achieve transparency and effectiveness of ICT for transport applications.

Moving forward, standardized specifications are essential to ensure better quality of VGI or crowdsourced datasets (Goodchild, 2008). Meanwhile the contribution checks and controls should be minimized in the field of crowdsourcing to prevent potential restrictions on data contributors (Girres and Touya, 2010; Goodchild and Glennon, 2010). Despite the point that accuracy of open data depends on its specific usage and purpose (Tauberer, 2012), access barriers should not impede the involvement of actual users in the process of generating such standards or specifications (Girres and Touya, 2010). To conclude, ICT for transport can enable diverse communities, reducing data collection and update costs. However, particular attention should be paid in the divergent contexts of developed and developing countries.

Table 6.1 Stakeholders' role in promoting the use of open data

	Current status	Future role
Citizen	– Mainly a consumer of open transport data via ICT – Participate in the process of crowdsourcing transport datasets – Help improve the quality of crowdsourced or open transport data by editing existing features continuously	– More effective commuting planning using ICT applications built based on open transport data despite the importance of geographic areas – Interactive with local and national government to jointly define the specifications of open geospatial and transport data – Improving data quality in real time when infrastructure is available
Local authority and national government	– Promoters of open transport data initiatives – Tend to focus on overcoming technical barriers in open data initiatives – Lower concern for remote areas or those with low ICT infrastructure	– Control and manage geolocation data – More emphasis put on increasing use and impact of open transport data initiatives – Integrate quality control and crowdsourcing data evaluation into existing processes – Foster citizen education through the release of existing open databases
International organization	– Support ICT and transport infrastructure construction in developing countries	– Raise awareness and promote usage of ICT for transport applications in developing countries – Facilitate quality improvement of open transport data through campaigns and social events – Help setting and implementing transport data quality standards
Private sector	– Facilitate crowdsourcing mechanism in collecting and sharing transport data – Initiate creation of common transport data standards – Collaborate with government and international organizations to promote ICT usage in rural areas in developing countries	– Fund innovative technology solutions to address transport issues in both developed and developing countries – Strengthen relationship with Open Data Initiative, by adopting open datasets in products design cycle – The role of private sector investment should be leveraged in transport sector in international development

ACKNOWLEDGEMENT

The authors would like to extend their gratitude to the following people whose support, input and collaboration were instrumental in the formation of this chapter: Holly Krambeck, Senior Transport Economist, Transport and ICT Global Practice; Dmitry Zaviralov, GIS Analyst, Information and Technology Solutions and the feedback received at the workshop hosted by the RSA Research Network on 'Using Information and Communication Technology to facilitate smart and green regional growth' during the American Association of Geographers (AAG) 2013 Annual Meeting.

REFERENCES

Antrim, A. and Barbeau, S. (2013) The many uses of GTFS data – opening the door to transit and multimodal applications, available at: http://www.locationaware.usf.edu/wp-content/uploads/2010/02/The-Many-Uses-of-GTFS-Data-%E2%80%93-ITS-America-submission-abbreviated.pdf.

Ather, A. (2009) A Quality Analysis of OpenStreetMap Data. MEng thesis, University College London, available at: ftp://ftp.cits.rncan.gc.ca/pub/cartonat/Reference/VGI/Dissertation-OpenStreepMap-Quality-Aather-2009.pdf.

Auer, R., Bizer, C., Kobilarov, G., Lehmann, J., Cyganiak, R. and Ives, Z. (2007) DBpedia: a nucleus for a web of open data. *Lecture Notes in Computer Science*, 4825: 722–35.

Brigham, C., Gilbert, S. and Xu, Q. (2013) Open geospatial data: an assessment of global boundary datasets, World Bank.

Chilton, S. (2009) Crowdsourcing is radically changing the geodata landscape: case study of OpenStreetMap. Paper presented at the 24th International Cartographic Conference, Chile.

Geenhuizen, M. (2009) ICT applications on the road to sustainable urban transport. *European Transport*, 41(2009): 47–61.

Girres, J. and Touya, G. (2010) Quality assessment of the French OpenStreetMap Dataset. *Transactions in GIS*, 14(4): 435–59.

Goodchild, M. (2008) Spatial accuracy 2.0. In Proceedings of the Eighth International Symposium on Spatial Data Accuracy Assessment in Natural Resources and Environmental Sciences, Shanghai, China.

Goodchild, M. and Glennon, J. (2010) Crowdsourcing geographic information for disaster response: a research frontier. *International Journal of Digital Earth*, 3(3): 231–41.

Goodchild, M., Fu, P. and Rich, P. (2007) Sharing geographic information: an assessment of the geospatial one-stop. *Annuals of the Association of American Geographers*, 97(2).

Google (2012) General Transit Feed Specification Reference. Available at: https://developers.google.com/transit/gtfs/reference?csw=1.

Graham, M. (2013) Geographies of information in Africa: Wikipedia and uneven user-generated content, R-Link, January–March, pp. 40–41.

Guptill, S. and Morrison, J. (eds) (1995) *Elements of Spatial Data Quality*. Oxford: Pergamon.

Haklay, M. (2010) How good is volunteered geographical information? A comparative study of OpenStreetMap and Ordnance Survey datasets. *Environment and Planning B: Planning and Design*, 37(4): 682–703.

Haklay, M. and Weber, P. (2008) OpenStreetMap: User-generated Street Maps. *Pervasive Computing*, 7(4): 12–18.

Halonen, I. (2012) Being open about data: analysis of the UK open data policies and applicability of open data. Available at: http://www.finnish-institute.org.uk/images/stories/pdf2012/being%20open%20about%20data.pdf.

Heipke, C. (2010) Crowdsourcing geospatial data. *SPRS Journal of Photogrammetry and Remote Sensing*, 65(6): 550–57.

Howe, J. (2006) The rise of crowdsourcing. *Wired Magazine*, 14(6): 1–4.

International Development Center of Japan (ICDJ) (2004) Transport and ICT. Making infrastructure pro-poor. POVNET 2nd Workshop on Infrastructure for Poverty Reduction. Final report 30 November.

Kresse, W. and Fadaie, K. (2003) *ISO Standards for Geographic Information*. Berlin, Springer-Verlag.

Lessig, L. (2007) 8 principles of open government data. Available at: http://www.opengovdata.org/home/8principles.

Mandel, B. and Ruffert, E. (2000) Generalised transport-data format (GTF): data, model and machine interaction. Paper presented at the In First ITEM Workshop, Montreal, Canada.

Miller, H. (2013) Beyond sharing: cultivating cooperative transport systems through geographic information science. *Journal of Transport Geography*, 31: 296–308.

Nielsen, O., Israelsen, T. and Nielsen, E. (1998) Bridges to GIS – Methodology, Deliverable D5 and D6. BRIDGES Contract No. PL96-1138. EU, DG7, 4th Framework Programme.

Oort, P. (2006) Spatial data quality: from description to application. Unpublished doctoral dissertation, Wageningen University.

Open Geospatial Consortium (2008) Geospatial data quality survey – open geospatial consortium data quality working group. Available at: https://www.google.com/url?sa=t&rct=j&q=&esrc=s&source=web&cd=2&cad=rja&ved=0CDMQFjAB&url=http%3A%2F%2Fportal.opengeospatial.org%2Ffiles%2F%3Fartifact_id%3D30415&ei=HpAuUqbfHebj4AOjkYDIBA&usg=AFQjCNHt0fBPsQOM2OyyNpVVXnUJ4xWJGw&sig2=ooAVsmvko3DsIYxVFVIr6g.

SKM, Ecological Associates and ACIL Tasman (2009) Spatial information in the New Zealand economy. Available at: https://www.google.com/url?sa=t&rct=j&q=&esrc=s&source=web&cd=1&cad=rja&ved=0CDAQFjAA&url=http%3A%2F%2Fwww.crcsi.com.au%2FDocuments%2Fspatial-information-in-the-new-zealand-economy-200.aspx&ei=c-FRUqn9HJK14AOlh4D4Ag&usg=AFQjCNG4_4dum2R6jQgqmUrbmSGdPI0yKg&sig2=6K1yCPJdE6u68epS2yMRlQ&bvm=bv.53537100,d.dmg.

Socrata, Sunlight Foundation, Personal Democracy Forum, GovLoop, Code for America and Eaves, D. (2011) Open government data benchmark study. Available at: http://www.socrata.com/benchmark-study.

Surowiecki, J. (2005) *The Wisdom of the Crowds: Why the Many are Smarter than the Few*, New York: Doubleday.

Tauberer, J. (2012) Open government data. Available at: http://opengovdata.io.

Teeuw, W., Koolwaaij, J. and Peddemors, A. (2012) User behaviour captured by mobile phones. In Wichert, R., Van Laerhoven, K. and Gelissen, J. (eds) *Constructing Ambient Intelligence*. Berlin and Heidelberg: Springer, pp. 81–90.
Thomopoulos, N. and Karanasios, S. (2014) Social impacts of information and communication technologies in transport. In Garrett, M. (ed.) *Encyclopedia of Transportation: Social Science and Policy*, Thousand Oaks: SAGE, pp. 575–8.
Universitas Gadjah Mada (2013) Evaluation of OpenStreetMap data in Indonesia. Available at: http://oo.or.id/2013-07-05/Dokumen%20Laporan/Final%20 Report-OSM%20Evaluation%20in%20Indonesia_2012.pdf.
Waze (2012) Waze launches in South Africa, 14 March. Available at: http://blog.waze.com.
Walpersdorf, A., Bouin, M., Bock, O. and Doerflinger, E. (2007) Assessment of GPS data for meteorological applications over Africa: study of error sources and analysis of positioning accuracy. *Journal of Atmospheric and Solar-terrestrial Physics*, 69(12): 1312–30.
Wikipedia (2013) OpenStreetMap Stats. Available at: http://wiki.openstreetmap.org/wiki/Stats.
Xia, J. (2012) Metrics to measure open geospatial data quality. *Science and Technology*, Winter.
Yahoo (2013) Waze sale signals new growth for Israeli high tech, Available at: http://news.yahoo.com/waze-sale-signals-growth-israeli-high-tech-174533585.html.
Zhao, Y. (2000) Mobile phone location determination and its impact on intelligent transport systems. *Intelligent Transport Systems*, 1(1): 55–64.
Zook, M., Graham, M., Shelton, T. and Gorman, S. (2010) Volunteered geographic information and crowdsourcing disaster relief: a case study of the Haitian earthquake. *World Medical & Health Policy*, 2(2): 7–33.

7. Impacts of vehicular communication networks on traffic dynamics and fuel efficiency

Wei Lu and Lee D. Han

1 INTRODUCTION

The rapid development of wireless communication and information technologies has pushed the industry of vehicular communication systems to improve the comfort, safety and operational efficiency of transport systems (Sichitiu and Kihl, 2008, Toor et al., 2008, Lu et al., 2009). Banister and Stead (2004) studied the influence of Information and Communication Technology (ICT) on transport from production, living and working perspectives. Hsinchun et al. (2004) explored the intelligence and security informatics for homeland-security-related applications using ICT and transport techniques. There are three areas summarizing the application of ICT for transport areas: operation of all modes and management of networks; transport information and guidance to the users; operation and management of freight systems (Giannopoulos, 2004). In this chapter, we place more emphasis on the operation and management of the transport network from a system-level perspective to provide theoretical and practical analysis of ICT for transport.

From a practical perspective, a wide range of projects about ICT for transport have been conducted since the 1980s (Papadimitratos et al., 2009, Karagiannis et al., 2011). The eSafety forum (Papadimitratos et al., 2006) is leading the European government public departments to cooperate with vehicle manufacturers. They focus on various aspects relating to intelligent vehicle safety issues. CarTALK2000 (Reichardt et al., 2002) focuses on the application of inter-vehicle communication in driving assistance systems. The main task is to evaluate the performance of driving assistance systems, design suitable data structures and algorithms for software development, and test its function in real world scenarios. Franz et al. (2001) promote FleetNet as the inter-vehicle communication platform, which implements Mobile Ad hoc Network (MANET) in vehicular communication networks. The Car-to-Car Communication Consortium (Eichler et al., 2006) is an open forum aiming at setting up a unique European standard for inter-vehicle communication, which consists of six vehicle manufacturers. Since 1986, the University of California, Berkeley and Caltrans have been cooperating to develop the California Partners for Advanced Transit and Highways (PATH) project. The purpose is to solve the problems of California's transportation systems through interdisciplinary knowledge, such as electrical engineering, information technology, economics and transport policy studies (Hedrick et al., 1994). Connected Vehicle Technology has been proposed to develop and deploy a fully connected transportation system. Most of these projects are government or research institute supported. Many of them only provide some prototype and simple test bed to verify various concepts of ICT for transport systems. It still needs uniform standards and commercialization tools for more efficient and effective applications.

From an academic perspective, vehicular communication networks have drawn considerable attention from researchers in transport engineering, electrical engineering, computer science, environmental science, economics, urban planning and other domains. Toor et al. (2008) summarize the applications and related technical issues of vehicular communication networks, whereas Wang and Lu explain in detail the inter-vehicle communication in a unidirectional traffic flow with and without dynamic traffic flow (Lu et al., 2009, Wang et al., 2009). Comparing the simulation results with the theoretical work of Jin and Recker (2010), Sun et al. researched different traffic scenarios on unidirectional traffic streams, bidirectional traffic streams, and multi-lane traffic streams (Sun et al., 2009, Lu et al., 2011, Sun et al., 2010). However, V2V (Vehicle-to-Vehicle) systems are restricted to the special mobility patterns of communication nodes in traffic networks, especially the dynamic traffic flow scenarios (Karnadi et al., 2007). The combination of traffic network and wireless communication network data is becoming an important part of vehicular communication systems. The realistic application of vehicular communication networks in car sharing systems is an important field in which to explore ICT for transport implications (Lu et al., 2013). It is important to understand the impacts of such mobility patterns on vehicular communication networks to develop better practical and efficient modern transport systems.

There are a limited number of researchers focusing on the environmental impact of information communication technologies specifically in transport systems. The Safe Road Trains for the Environment (SARTRE) project studies the environmental, safety and comfort benefits of vehicle platoons on the normal highway (Robinson et al., 2010). The Partners for

Advanced Transportation Technology (PATH) at UC Berkeley discussed the platooning effect on wind drag and fuel consumption (Michaelian and Browand, 2000). It indicated a 20–30 percent reduction in the coefficient of wind drag and a 10 percent reduction in fuel consumption. Bergenhem et al. (2012) presented performance measurements of a first prototype of the Vehicle-to-Vehicle system and provided supports for the SARTRE project. The US Environmental Protection Agency provides two main tools for vehicle fuel consumption and emission analysis: Physical Emission Rate Estimator (PERE) and Motor Vehicle Emission Simulator (MOVES) (Koupal et al., 2002). Since there is insufficient real world vehicle classification data, we mainly focus on the fuel consumption analysis with PERE in this chapter, which also provides direct reflections of environmental impacts.

The rest of this chapter is organized as follows. In section 2, we discuss the modeling process of vehicular communication networks in both vehicle platooning and dynamic traffic scenarios. Also the experimental set-up of the fuel consumption study in vehicle platooning is explained. In section 3, the simulation results of vehicular communication in different network configurations and traffic dynamic conditions are summarized. Also, the impact of vehicle platooning on fuel consumption is pointed out. Suggestions and recommendations based on these simulation results are provided for model developers and practitioners. In section 4, three pressing steps for researchers and practitioners and three further challenges in ICT for transport research are highlighted. Finally, some technical issues on modeling the development and simulation of such V2V applications are included in the Appendix.

2 METHODOLOGY

2.1 Mobility Model for Two Traffic Flow Scenarios

One of the benefits of implementing vehicular communication is that a vehicle platoon can be generated through autonomous driving technology. A vehicle platoon is a group of vehicles moving with the same speed and minimum space with the capability to accelerate or brake simultaneously. From a traffic flow perspective, vehicle platooning also has a uniform traffic flow. Figure 7.1 describes a simple vehicular communication network where five vehicles communicate with each other to form a vehicle-to-vehicle (V2V) communication network. Vehicle-to-Roadside (V2R) is also another kind of vehicular communication system. A vehicle platoon can be seen in Figure 7.1 as the three vehicles

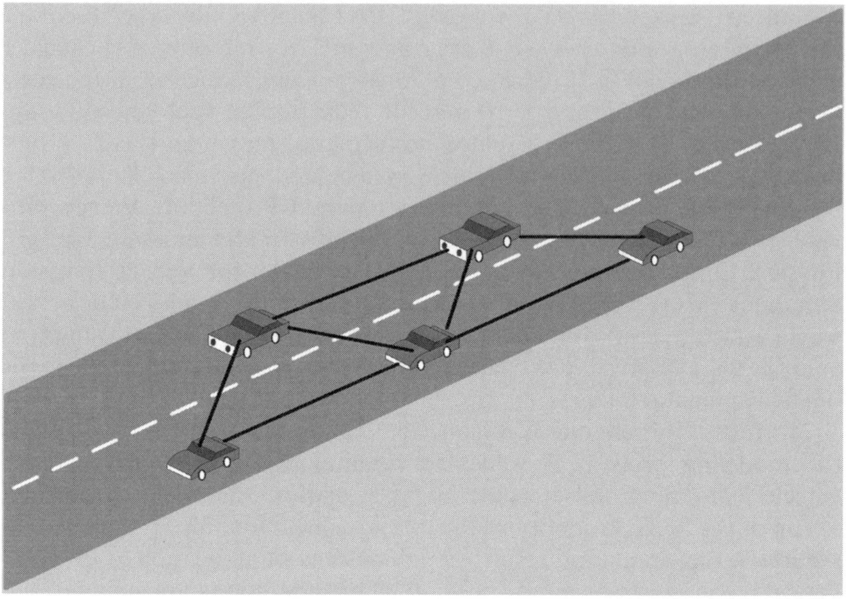

Figure 7.1 Vehicular communication network (vehicle-to-vehicle communication)

in the right-hand lane are moving with the same pattern. Since this chapter focuses on vehicle platooning and traffic dynamics, we only study V2V impacts here.

Due to the limitations of real-world vehicular communication network data and the high cost of experimenting with large-scale vehicular communication networks, we use modeling and simulation as our main methods to study the vehicular communication performance on traffic dynamics and fuel efficiency of vehicle platooning. We adopt Simulation of Urban Mobility (SUMO) as our traffic generation tool, which is an open-source, microscopic, multiple-model based traffic simulation software. The purpose of SUMO is to make the survey results more comparable and to make various algorithms in traffic flow easier to implement (Krajzewicz et al., 2002). Theoretical analysis of the mobility modeling of these two traffic flow scenarios is summarized in the Appendix (section A.1).

To adjust the impacts of randomness of computer simulation, a customized Monte Carlo simulation method is implemented in this chapter. The benefit of Monte Carlo simulation is that it provides more accurate and predicable simulation results for analysis. Details about the Monte Carlo model can be found in the Appendix (section A.2B).

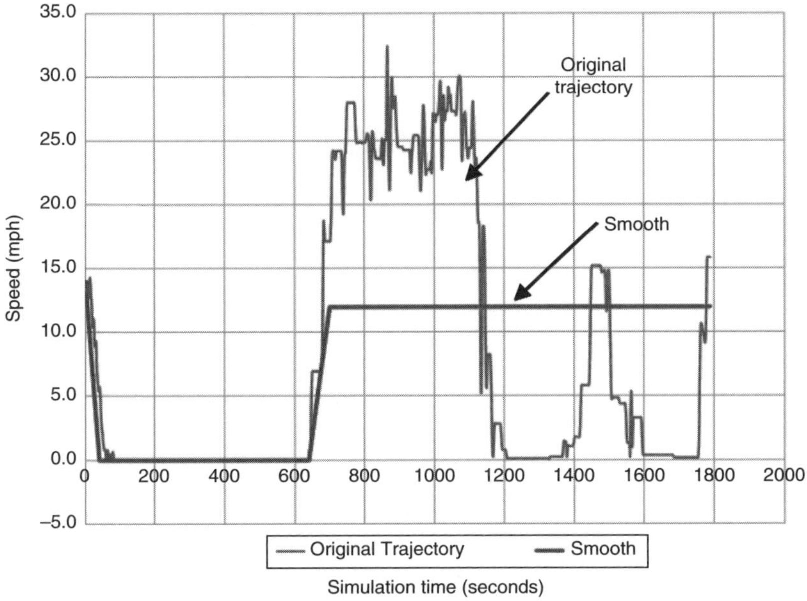

Figure 7.2 Smoothing with stop-and-start cycle

2.2 Fuel Consumption Experiment with Vehicle Platoons

This chapter estimates the changes in fuel consumption from the adoption of cooperative platooning. The real-world vehicle trajectory data on Peachtree Street, Atlanta, Georgia from Next Generation Simulation (NGSIM) is used as the input for PERE, including motor characteristics, fuel type, vehicle characteristics and drive cycle. The output of PERE is fuel consumption data.

In this chapter, one car case is studied as the first step to analyze the fuel consumption. The vehicle trajectory is smoothed after vehicle platooning is adopted. Two smoothed versions of real trajectory data are created. The first trajectory data has a stop-and-start cycle, as Figure 7.2 shows. The unsmoothed line is the original trajectory and the bold smoothed one is the one with vehicle platooning. This case is more suitable when vehicle platoons approach an intersection. The second trajectory data has no stop-and-start cycle but has two different throughputs, as shown in Figure 7.3. The lower (V=7.5) line has the same throughput equal to the average speed of the original trajectory. The upper line (V=20) has higher throughput equal to the average rolling speed of the original trajectory, which is highly anticipated after adopting vehicle platooning.

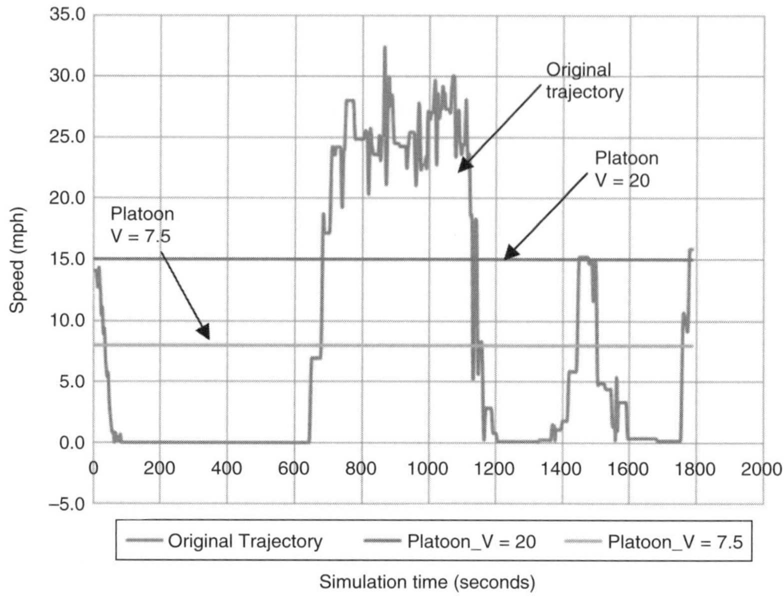

Figure 7.3 Smoothing with no stop-and-start cycles

3 EXPERIMENT EVALUATION

3.1 Simulation Results in Vehicle Platooning

For the network simulation, we use Network Simulator version 2 as our tool to simulate the vehicular communications in different vehicle mobility scenarios. The detailed network configuration is listed in the Appendix (section A.3).

To measure the network performance, two important parameters are used in this chapter: throughput and end-to-end delay. The throughput is the number of information packages successfully received during a certain time period. It determines how far and how successfully a particular vehicular communication can reach. The end-to-end delay is the average delay for information package transmission. It indicates how fast the information can be delivered in vehicular communication networks. One key comparison variable for different network scenarios is the transmission range. This is the furthest distance that the vehicular communication equipment can reach. The communication equipment always has limitations regarding transmission distance due to technology and cost limitations. This will provide a direction to find the best practical standards for

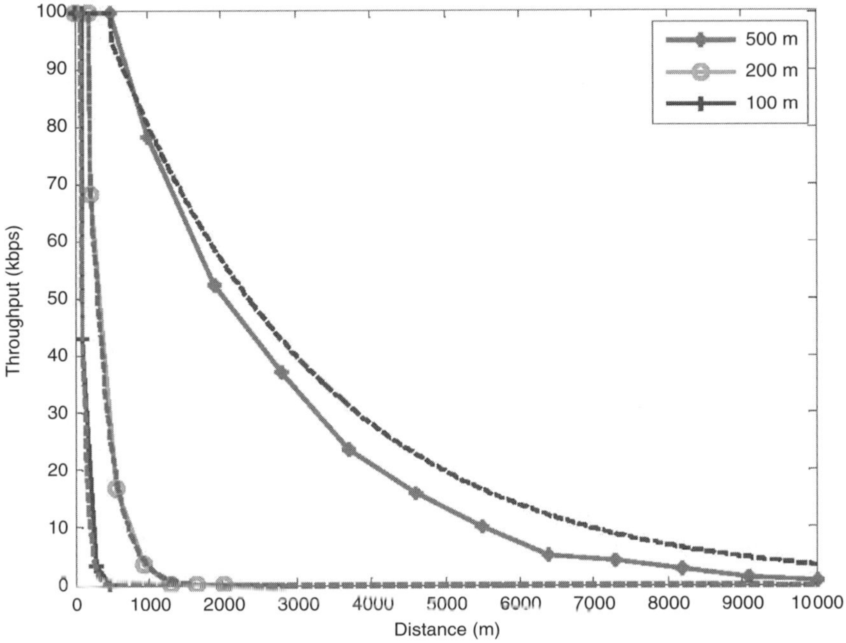

Note: The corresponding dashed lines are the values from the connectivity model.

Figure 7.4 Average throughput of a receiver for different transmission ranges

producing vehicular communication equipment. We run Monte Carlo simulations 500 times to get the average values for different transmission ranges γ = 500m, 200m and 100m.

Figure 7.4 shows that the throughput increases with transmission ranges for multi-hop connections in vehicle platooning scenarios. The dashed lines are theoretical values for communication performance in uniform traffic flow. Our practical simulation results are drawn with solid lines. They follow the theoretical trends on performance but have a weaker performance on a longer transmission range. After analyzing the network simulator results, low throughput appears to be the outcome of no successful communication chains, no available route and no buffer space in interface queue in the network. These results indicate that the vehicular communication does not work well when equipped vehicles are far from each other. The equipment with wider transmission capacity has better performance. Practitioners and policy makers should direct their attention towards the adoption of high standards on vehicular communication

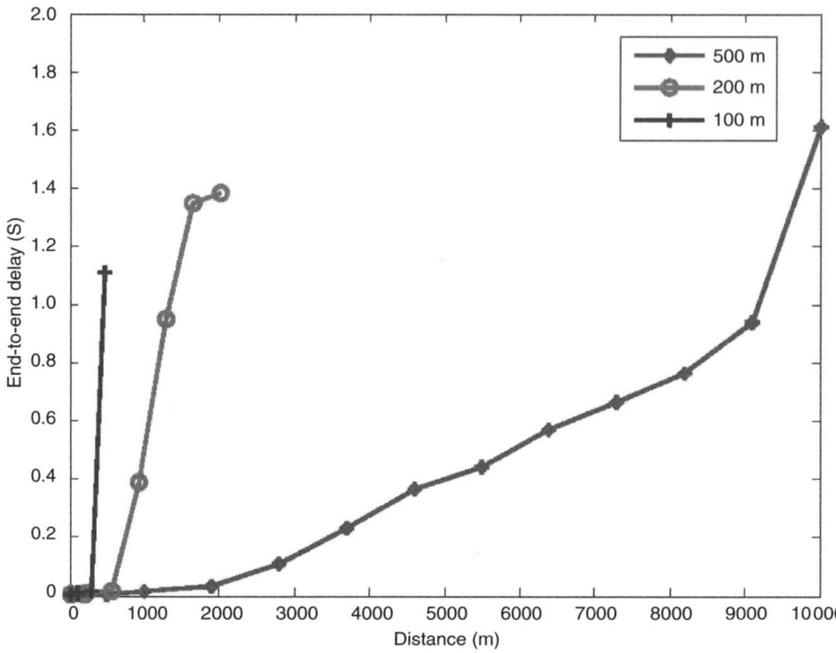

Figure 7.5 End-to-end delay between the source and the destination vehicles for different transmission ranges

equipment manufacturing (with high transmission capacity) and towards increased use of vehicular communication networks in dense urban areas to improve connectivity performance.

The end-to-end delay is another criterion used to analyze the performance. The less delay, the more the efficiency of the V2V system increases. The end-to-end delay increases when the distance between source and destination vehicles increases (Figure 7.5). For the same source and destination vehicles, a small transmission range increases the end-to-end delay. This is because the information package needs more hops in a small transmission range situation. This also proves that equipment with wider transmission capacity has a better performance from an information delay perspective.

3.2 Simulation Results in Dynamic Traffic Flow

Uniform traffic flow, such as a vehicle platoon, can only exist in a small group of vehicles. During the initial stage of applying autonomous driving

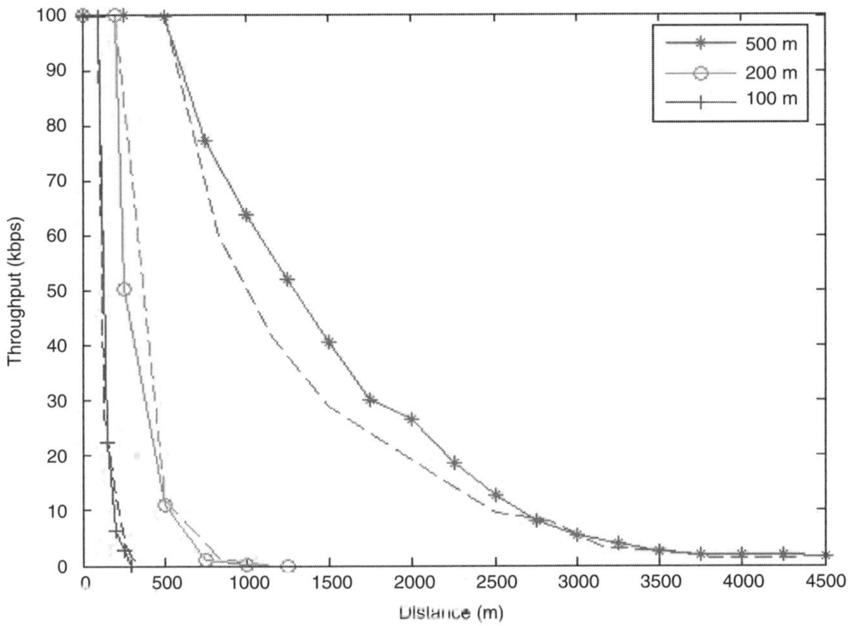

Note: The corresponding dashed lines are the values from the uniform scenario.

Figure 7.6 Average throughput of a receiver for different transmission ranges

vehicles, transport planners and decision makers must consider the reality of a mixed traffic flow of V2V equipped and unequipped vehicles. In dynamic traffic flow conditions, the speed and density of upstream traffic gradually changes into the downstream traffic situation. Figure 7.6 indicates that the throughput of the same destination vehicle will increase with a higher transmission range for both the uniform scenario and the dynamic scenario. There is hardly any influence on the throughput of the destination vehicle with dynamic change of vehicles' topology as the transmission range is 100m or 200m. With the low penetration rate and short transmission range, there is less possibility for disconnection of the communication networks. When the transmission range is 500m, the throughput at the shockwave scenario is higher than the value of the uniform scenario. With the introduction of high transmission equipment, the vehicular communication has slightly better performance due to the random distance between equipped vehicles. This helps the planners to ease concerns about vehicular communication performance in mixed dynamic traffic flow scenarios.

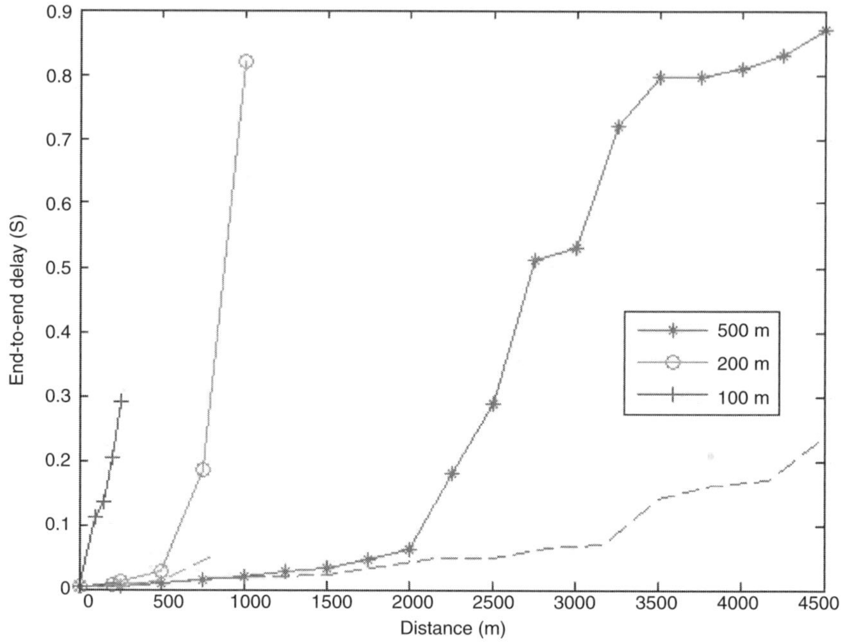

Note: The corresponding dashed lines are values from the uniform scenario.

Figure 7.7 End-to-end delay between the source and the destination vehicles for different transmission ranges

Figure 7.7 compares the end-to-end delay of both the uniform and shockwave scenarios. Simulation findings demonstrate that the end-to-end delay between the same source vehicle and destination vehicle decreases with the increase of transmission range. On the other hand, the end-to-end delay of the shockwave scenario is remarkably longer than that of the uniform scenario. If the density of upstream traffic becomes the status of downstream traffic, this will increase the processing time between source vehicle and forward vehicles and then lead to a significantly augmented value of end-to-end delay. Under the same network configuration, the dynamic traffic flow increases vehicular communication performance. This makes the vehicular communication network easier to adapt in existing normal traffic conditions.

Table 7.1 Fuel consumption comparison

Results	Original	W/ Stops	Smooth (8 mph)	Smooth (15 mph)
Fuel economy (mpg)	15.33	14.96	17.48	20.88
Delta% mpg	N/A	−2.41	14.02	36.22

3.3 Fuel Consumption Analysis with different scenarios

The fuel consumption of a given vehicle with platooning is analyzed in Table 7.1. The fuel economy measurement, that is, miles per gallon (mpg), is adopted to evaluate performance. Four different scenarios are compared, including the following: original trajectory; with stop-and-start smoothed trajectory; without stop-and-start at average 8 mph smoothed trajectory; and without stop-and-start at average 15 mph smoothed trajectory. The mpg value of the original unsmoothed trajectory is 15.33. After smoothing with stop and start at the intersection case, the fuel consumption was decreased by 2.41 percent. When vehicles stop at an intersection with the engine still running, they consume more fuel for just waiting without moving towards their destination. In total, the mpg value is lower than an unsmoothed real trajectory. The 8 mph smoothed trajectory case improved the fuel consumption by 14.02 percent and the 15 mph smoothed trajectory case improved the fuel consumption further by 36.22 percent. This analysis indicates that faster average moving speed decreases the travel time and increases the fuel efficiency, which is an important finding about ICT for transport with policy implications for real world implementation.

4 CONCLUSIONS

In this chapter, we studied the performance of a special vehicle-to-vehicle communication network in both uniform and dynamic traffic flow scenarios. Also, we conducted a fuel consumption analysis for a vehicle platooning situation. Based on the work of Wang et al. (2009), we used throughput and end-to-end delay as the performance measures to illustrate how far and how fast the vehicular communication network works. Through these analyses and discussions, three V2V recommendations arise based on the simulation results, which are of interest for researchers and practitioners.

- Communication equipment selection. If sufficient funding exists, it is more effective to use communication equipment with a higher transmission range. No matter if it is in uniform traffic flow conditions

(such as a freeway in a rural area) or in dynamic traffic flow conditions (such as major roads within urban areas), the equipment with a high transmission range can improve the communication quality and decrease the impacts of transmission delay, which would aid in meeting wider transport policy targets.
- Penetration rate study. In the near future, we cannot expect that every vehicle will have the technology to communicate with all other vehicles. In fact, we do not envisage all vehicles to be V2V equipped to have optimal traffic conditions. What proportion of vehicles should be equipped with communication technology becomes a pertinent research topic, not only from a funding perspective, but also from a technical perspective.
- Environmental analysis. In this chapter, only fuel consumption in PERE is considered to study the impact of vehicle platooning based on vehicle-to-vehicle (V2V) communication technology. To study the benefits of vehicular communication technology, a comprehensive analysis with MOVE could produce more environmental performance data, including air quality and emission efficiency, improving the robustness of this analysis.

Moreover, three key issues are stressed in relation to the challenges and opportunities regarding the use of ICT for transport:

- Safety. The major concern about adopting vehicular communication technology is driver and passenger safety. Vehicle platooning requires high accuracy in vehicle electrical and mechanical systems. The existing test cases can only handle low-speed vehicles. This actually decreases the transport efficiency by reducing system performance. To achieve vehicle platooning with free-flow speed, a series of experiments and real-world testing are needed.
- Legislation. Although the technology for autonomous vehicles nowadays (such as the Google driverless car) is much more mature than before, users are still concerned about such new technological advancements. Until now, only four states in the United States (Nevada, Florida, California and Michigan) and the United Kingdom have begun to permit testing of autonomous vehicles on public roads. Real world testing on roads is a key factor for the success of such ICT for transport applications.
- Cost. Most vehicles with vehicular communication capability are still in the prototype stage. The cost of producing one autonomous vehicle is much higher than the best Tesla electrical vehicle. Since electrical vehicle technology is more mature than autonomous

vehicles, it will be highly desirable to find a way to integrate the vehicular communication technologies in the electric vehicles to provide a smart, compatible and sustainable transport system in the near future.

ACKNOWLEDGEMENT

This work is supported by the ORNL Joint Directed Research and Development project and the National Natural Science Foundation of China (60974092). Views expressed in this chapter are those of the authors and do not reflect the views of Oak Ridge National Laboratory (ORNL).

REFERENCES

Banister, D. and Stead, D. 2004. Impact of information and communications technology on transport. *Transport Reviews*, 24, 611–32.

Bergenhem, C., Hedin, E. and Skarin, D. 2012. Vehicle-to-vehicle communication for a platooning system. *Procedia – Social and Behavioral Sciences*, 48, 1222–33.

Eichler, S., Schroth, C. and Eberspächer, J. 2006. Car-to-car communication. Proceedings of the VDE-Kongress. Aachen: VDE Verlag GmbH.

Franz, W., Eberhardt, R. and Luckenbach, T. 2001. Fleetnet – Internet on the road. Proceedings of the 8th World Congress on Intelligent Transport Systems, Sydney, October.

Giannopoulos, G.A. 2004. The application of information and communication technologies in transport. *European Journal of Operational Research*, 152, 302–20.

Hedrick, J.K., Tomizuka, M. and Varaiya, P. 1994. Control issues in automated highway systems. *IEEE Control Systems*, 14, 21–32.

Hsinchun, C., Fei-Yue, W. and Daniel, Z. 2004. Intelligence and security informatics for homeland security: information, communication, and transportation. *IEEE Transactions on Intelligent Transportation Systems*, 5, 329–41.

Jin, W.L. and Recker, W. 2010. An analytical model of multihop connectivity of inter-vehicle communication systems. *IEEE Transactions on Wireless Communications*, 9, 106–12.

Jin, W.L. and Zhang, H.M. 2004. A multicommodity kinematic wave simulation model of network traffic flow. *Transportation Research Record: Journal of the Transportation Research Board*, 1883, 59–67.

Karagiannis, G., Altintas, O., Ekici, E., Heijenk, G., Jarupan, B., Lin, K. and Weil, T. 2011. Vehicular networking: a survey and tutorial on requirements, architectures, challenges, standards and solutions. *IEEE Communications Surveys & Tutorials*, 13, 584–616.

Karnadi, F.K., Mo, Z.H. and Lan, K. 2007. Rapid generation of realistic mobility models for VANET. Proceedings of the IEEE Wireless Communications and Networking Conference (WCNC), Hong Kong, 11–15 March, pp. 2506–11.

Khaled, Y., Ducourthial, B. and Shawky, M. 2005. IEEE 802.11 performances for inter-vehicle communication networks. *Proceedings of the 61st IEEE Vehicular Technology Conference, Spring, Vols 1–5*, pp. 2925–29.

Koupal, J., Michaels, H., Cumberworth, M., Bailey, C. and Brzezinski, D. 2002. EPA's plan for MOVES: a comprehensive mobile source emissions model. Proceedings of the 12th CRC On-Road Vehicle Emissions Workshop, San Diego, CA, 15–17 April.

Krajzewicz, D., Hertkorn, G., Wagner, P. and Rossel, C. 2002. An example of microscopic car models validation using the open source traffic simulation SUMO. Proceedings of Simulation in Industry, 14th European Simulation Symposium, pp. 318–22.

Lu, W., Bao, Y.L. and Bai, H. 2011. Research on Inter-vehicle Communication Performance in Unidirectional Traffic Flow. *Jisuanji Gongcheng/Computer Engineering*, 37.

Lu, W., Han, L. and Cherry, C. 2013. Evaluation of vehicular communication networks in a car sharing system. *International Journal of Intelligent Transportation Systems Research*, 11, 113–19.

Lu, W., Bao, Y., Sun, X. and Wang, Z. 2009. Performance evaluation of inter-vehicle communication in a unidirectional dynamic traffic flow with shockwave. Proceedings of ICUMT'09, IEEE International Conference on Ultra Modern Telecommunications & Workshops, pp. 1–6.

Michaelian, M. and Browand, F. 2000. Field experiments demonstrate fuel savings for close-following, California PATH research report, Institute of Transportation Studies, University of California at Berkeley.

Papadimitratos, P., Kung, A., Hubaux, J-P. and Kargl, F. 2006. Privacy and identity management for vehicular communication systems: a position paper. Paper prepared for workshop on standards for privacy in user-centric identity management, July, Zurich.

Papadimitratos, P., La Fortelle, A., Evenssen, K., Brignolo, R. and Cosenza, S. 2009. Vehicular communication systems: enabling technologies, applications, and future outlook on intelligent transportation. *IEEE Communications Magazine*, 47, 84–95.

Reichardt, D., Miglietta, M., Moretti, L., Morsink, P. and Schulz, W. 2002. CarTALK 2000: safe and comfortable driving based upon inter-vehicle-communication. Proceedings of the IEEE Intelligent Vehicles Symposium, pp. 545–50.

Robinson, T., Chan, E. and Coelingh, E. 2010. Operating platoons on public motorways: an introduction to the SARTRE platooning programme. Proceedings of the 17th World Congress on Intelligent Transport Systems, Busan, Korea, 25–29 October, pp. 1–11.

Sichitiu, M.L. and Kihl, M. 2008. Inter-vehicle communication systems: a survey. *IEEE Communications Surveys & Tutorials*, 10, 88–105.

Sun, X., Bao, Y., Dai, J., Lu, W. and Gao, R. 2009. Performance analysis of inter-vehicle communications in multilane traffic streams. Proceedings of the IEEE Vehicular Networking Conference (VNC), Tokyo, pp. 1–6.

Sun, X., Bao, Y., Lu, W., Dai, J. and Wang, Z.A. 2010. Study on performance of inter-vehicle communications in bidirectional traffic streams. Proceedings of the Second International Conference on Future Networks, Sanya, China, pp. 68–72.

Toor, Y., Muhlethaler, P. and Laouiti, A. 2008. Vehicle ad hoc networks:

applications and related technical issues. *IEEE Communications Surveys & Tutorials*, 10, 74–88.
Wang, Z., Lu, W. and Jin, W. 2009. A study on information throughput of inter-vehicle communications in a unidirectional traffic stream. *Proceedings of the WRI International Conference on Communications and Mobile Computing*, IEEE, pp. 396–401.

APPENDIX

A.1 Mobility Model for Two Traffic Flow Scenarios

The classic microscopic traffic theory, the car following model, can be used to calculate the trajectory of each car. The model is:

$$\begin{cases} \frac{d^2}{dt^2}x_{i+1}(t + T) = \lambda\frac{d}{dt}s_i(t) \\ s_i(t) = x_i(t) - x_{i+1}(t) \end{cases} \tag{A.1}$$

Where s_i (t) is the distance between two adjacent vehicles; x_i (t) is the position of vehicle i in the special coordinates; λ is the sensitive coefficient.

In the real traffic flow, the behavior of the traffic flow is similar to the behavior of the fluid wave (Jin and Zhang, 2004). We assume that there are two adjacent areas (A and B) with different traffic density ρ_u and ρ_d. The interface of these two areas is called wave front S and the speed of S is v_s. The respective vehicle speeds in areas A and B are v_u and v_d. We can calculate the number of vehicles (N) passing by the interface S within the time t as follows:

$$N = (v_u - v_s)\rho_u t = (v_d - v_s)\rho_d t \tag{A.2}$$

Then we can obtain the speed of the traffic wave as below:

$$v_s = \frac{q_d - q_u}{\rho_d - \rho_u} \tag{A.3}$$

The traffic wave describes the converting of two different traffic conditions, where v_s represents the direction and process of converting. When $v_s > 0$, the direction of traffic wave is the same as that of the traffic flow; when $v_s < 0$, the direction of the traffic wave is the opposite to that of the traffic flow; when $v_s = 0$, the traffic wave doesn't move.

A.2 Monte Carlo Simulation Model

The information relay from the source to the destination is a random process and the throughput and end-to-end delay for this communication link depends on the connectivity of the V2V based on the value of transmission ranges. To evaluate the performance of a communication link within V2V, we use Monte Carlo simulations through repeated, random realizations of Bernoulli trials. For a dynamic traffic flow with K+1 vehicles, the flow-chart of Monte Carlo simulations of V2V is shown as Figure 7A.1, in which each step is carried out as follows.

Inputs for the experiments include market penetration rate μ, transmission range γ, and the number of simulation times M.

In each experiment, we use $K + 1$ uniformly distributed random variables to determine which vehicles are equipped. Here the random variable X_k, corresponding to vehicle k ($k = 0, \ldots, K$), is generated in [0, 1]. If $X_k \leq \mu$, vehicle k is equipped, and we can obtain a Bernoulli trial.

For each Bernoulli trial, we provide trajectories of all equipped vehicles to NS-2 and simulate V2V from the sender. For experiment m, the throughput is denoted by T_i ($i = 1, \ldots, M$). If it cannot set up a communication path from the sender to the receiver, the throughput is 0 kbps. Likewise, the end-to-end delay is denoted by D_i.

After the M experiments, the throughput and end-to-end delay of an equipped vehicle at x can be computed by:

$$T_M = \frac{\sum_i T_i}{M}, T_i = \frac{\sum_j P_j}{t_{\max} - t_0}, j = 0,1,2 \ldots, \max \tag{A.4}$$

$$D_M = \frac{\sum_i D_i}{M}, D_i = \frac{\sum_j d_j}{\max + 1}, j = 0,1,2 \ldots, \max \tag{A.5}$$

A.3 Simulation Configuration

The simulation of the inter-vehicle communication under NS-2 environment is configured according to Khaled et al. (2005). We use constant bit rate (CBR) generators for the application layer. For each Bernoulli trial, the simulation time is 180s because the time when the first vehicle encounters a shockwave is 166.15s. User Datagram Protocol (UDP) is used for the transport layer and Ad hoc On-Demand Distance Vector (AODV) routing protocol is selected for the network layer. At the Media Access Control (MAC) layer and the physical layer, IEEE 802.11 DCF and the two-ray ground model are used for the study.

For the traffic networks, we consider information propagation in a shockwave dynamic traffic flow on a four-lane road-way. The probability for vehicles to be equipped equals the market penetration rate μ = 10%. The interval for sending packets is 0.02s and the packet size is 230 Bytes.

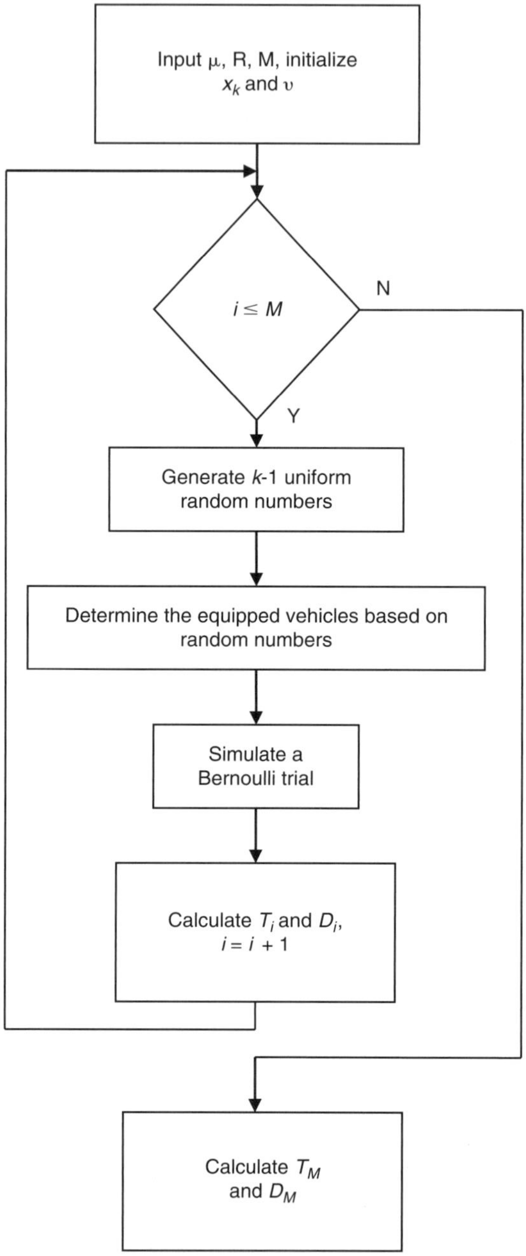

Figure 7A.1 A Monte Carlo simulation model of inter-vehicle communication in a shockwave dynamic traffic flow

8. ICT and planning for increased cycling in Europe – now and tomorrow

Pelle Envall

1 INTRODUCTION

CyCity has made use of ICT in the form of data collected with GPS devices since 2009. Few people then owned a smartphone and even fewer had tested a GPS function on a mobile phone, even if their phone had one. During 2010–12 CyCity tested different types of GPS devices and their abilities to correctly record the position of a cyclist in an urban environment. GPS data was collected, first in some small pilot schemes, then on a larger scale. The GPS data was saved in a GIS database created to the project's specification. The CyCity project then started to explore data quality and the ability of GPS data to provide useful information for cycle planning. Soon it was acknowledged that in order to make intelligible use of cyclists' GPS data the project needed to develop new software for data exploration, analysis and visualization. The starting point for the project in 2009 was *not* how ICT could make cycling more attractive. The CyCity starting point was something slightly different, namely the fact that many people love the freedom of riding a bicycle and that comfort and travel time varies from city to city and from neighbourhood to neighbourhood, a fact that is poorly appreciated in commonly applied software tools for transport planning. Hence, the CyCity starting point was how currently available ICT in the form of transport analysis software can be improved to better accommodate the needs and preferences of cycle users.

2 EXPLORING THE NEED FOR BETTER ICT APPLICATIONS

2.1 GPS Technology and its Opportunities

Overview

GPS devices provide information on where a particular device (vehicle) is positioned at a given point in time. Devices that are linked to a specific user (a person on a bicycle) can potentially provide planners with a number of interesting characteristics about travel behaviour and transport networks, including:

- cycling speed;
- location and lengths of delays;
- route choices;
- demographic data such as differences in route choice or cycling speed depending on age, gender or type of street design.

General opportunities within transport planning

The studies initially identified two main links with transport planning where cyclist GPS data may prove useful for improving transport and urban planning practice:

- First, GPS devices can potentially prove useful in reducing data collection costs.
- Second, GPS devices may provide new data of better quality and by means of this increase our understanding of cycling behaviour as well as improve awareness regarding the factors that underpin people's modal choice. This in turn may aid in meeting transport planning goals, such as increasing the cycling modal split or reducing cycling fatalities.

Data about transport mode as well as cycling flows and user behaviour are vital parts of cost-effective transport and urban planning. But data collection is itself a cost in human and financial resources. Transport models detailing vehicle flows and delays are more or less a mainstream feature within transport planning in many cities in the Western world. These models have been and still are underpinned by collection of traffic flow data. There has been a continuous development of more sophisticated data collection methodologies including floating car and vehicle number plate recognition systems to survey origin and destinations, automatic counters and more. However, the collection of cycling data is considerably

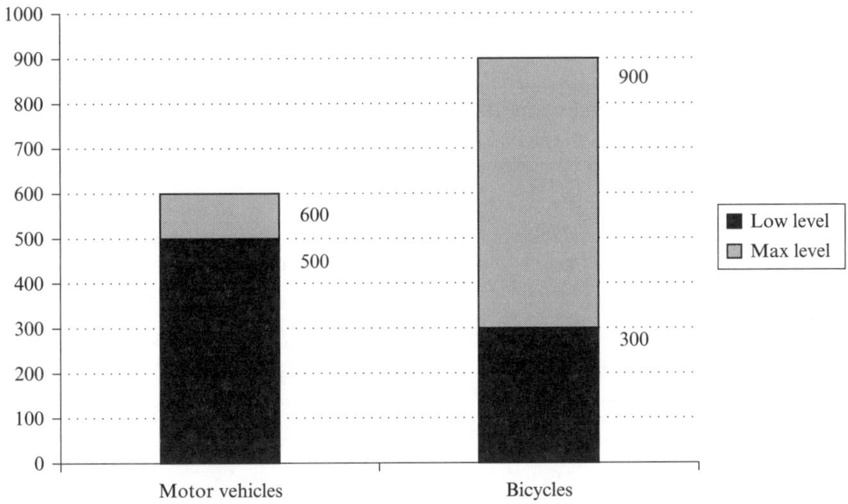

Note: Maximum level for bicycles is weekday mornings April to October.

Source: WSP (2012).

Figure 8.1 Estimate of peak hour flows of cyclists and motor vehicles per direction on Götgatan in central Stockholm

less developed. One way of viewing this is the fact that there are very few opportunities to include cycling data in common transport model software analysis, not even when cycling is the main mode during rush hour, as on streets in Stockholm. See Figure 8.1 for an example.

Opportunities within objective-led transport planning

There appears to be growing interest about objective-led planning focusing on the achievement of quantifiable and tailor-made targets within urban transport planning. Literature indicates that Western cities have multiple policy objectives for transport planning (for example Marsden et al., 2010). Planning advice commonly recommends setting up quantitative performance targets (for example EC, 2007), even if this also has some potential disadvantages (May et al., 2003). One such measurable goal that Swedish and other European local and regional authorities commonly set is to increase the modal share of cycling (Brundell-Freij and Envall, 2012; Sloman et al., 2009). As shown in Table 8.1, the levels of increase in the modal share of cycling that Swedish cities have set as their policy target were quite often high, sometimes aiming for entering the world's golden

Table 8.1 Goals to increase cycling in selected Swedish cities

City	Approximate cycling modal split	Target for cycling modal split	Year when target should be reached
City of Gothenburg	9%	12%	2010
City of Malmö	23%	Walking, cycling and public transport modal split should increase from 59% to 67%	2030
Stockholm City Council	8%	Twice as many cyclists at places where cycle flows are counted	2015
Linköping local authority	30%	40%	2028

Source: Niska et al. (2010); Brundell-Freij and Envall (2012).

league of cycling cities, that is, those with a cycling modal split of well above 30 per cent.

In order to meet the goal of increasing cycling it is important to identify the following:

- What policy measures and changes in the transport system are linked to increasing the cycling modal split?
- What is the current state of urban transport networks?
- Is the network sufficiently cycle friendly? Are there gaps in the network that make people apprehensive about cycling?

User perception of improved decision support

A survey of 235 planners responsible for and involved in transport and cycle planning in Sweden indicated a number of interesting findings (Envall, 2009). Amongst other things, the study identified a relatively strong demand for improving decision support for cycle planning in Sweden. The majority of respondents (54 per cent) indicated that better decision support on a national and regional level is *important* or *very important* in order to make it easier to implement better cycle facilities. The above-mentioned baseline study also showed that decision support for local cycle planning at the time was out of date, raising questions about potential implementation gaps.

Measuring cycling quality is a key opportunity
In order to know how to increase the modal share of cycling and improve urban transport efficiency generally, it is important to assess the quality of the street network from a cycling point of view. GPS devices provide an opportunity to do this by collecting data on cyclists' travel speed and transmitting it through ICT. Travel speed is information not commonly available for cycling today (but frequently available for car-oriented planning). Data on cycling behaviour can then be used to inform the need for, and analyse the usefulness of, different policy measures and transport planning at a strategic level. GPS devices can also provide data on route choices. Route choice data for cyclists can potentially be used to pinpoint user preferences, either by identifying high-quality routes that attract users to take a detour, or by identifying road links that are not perceived of adequate level for cycling and therefore avoided by many users or specific user groups such as children, women or less experienced cyclists.

2.2 Threats and Difficulties

The CyCity project identified two potential key threats or difficulties for using new GPS technology in the cycling planning field. Both key threats are non-technical. The first threat is that the contemporary strong interest globally about cycling-friendly urban policies may fade away if the practice of transport planning and its tools are not implemented without making use of the new opportunities such as data collected by means of GPS devices. Car-oriented urban planning has had a strong hold over resources for planning and urban design in Stockholm (for example Lundin, 2008) as well as in many other cities. One consequence of this is that resources and innovation when it comes to planning for cycling have been relatively scarce and ad hoc, as illustrated in the Stockholm case (Emanuel, 2012). However, bike sales reportedly overtook car sales in Europe in 2012 (BBC, 2013). In Sweden, bike sales clearly outnumbered car sales, which is a significant observation despite any weather-imposed restrictions on cycling. Hence, this threat related to lack of interest in cycling may seem less important at this point in time, at least in a Swedish and European context, which is particularly relevant when the ultimate aim is the transformation to a low carbon economy (Floater et al., 2013; NCE, 2014). A second and perhaps more vital threat or difficulty in many places is a perception of cycle planning (amongst planners, traffic engineers and mid-level management) as a low-cost and low-value service. This is a perspective that may effectively hinder the development and use of more sophisticated planning approaches and methodologies. It could be anticipated that applying new technologies solely for cycle planning would make transport planning more costly than current ad hoc-oriented cycle planning practices. However, it could also be argued that more sophisticated planning approaches using better data would be more cost-efficient, better at identifying real user needs and targeting resources at measures that achieve policy goals effectively. This need has been identified in London, for example, while designing the cycle superhighways and looking for scheme sponsors for the second phase of the city-wide cycle sharing scheme.

There are of course other more technical difficulties as well. One such difficulty is the availability of appropriate GIS data on cycle networks. However, the CyCities experience from the two cities where GPS data has been collected is that it is possible to overcome this problem.

Policy interest in cycling: the Swedish context

Elected leaders in many Swedish cities and regions as well as in the national parliament (Riksdagen) pursue, as mentioned earlier, cycle-friendly planning policies in the form of high-level goals of increases in the modal split of cycling. However, several transport forecasts indicate that Sweden overall may experience a decline in cycling modal share over the coming years (for example Trafikverket, 2012, p. 19). The UK National Transport Model provides similar results (DfT, 2012, pp. 37–8). There is an obvious mismatch between the national forecasts and the ambitions of many local councils, in Sweden as well as in other European countries. But what is the source of this deviation? Is the national forecast misleading, not taking in local cycle network design initiatives, policy and behavioural shifts? Are local transport planners naive in pencilling in goals of increased cycling in transport plans without the measures and policy instruments in place to achieve this? Do national transport authorities pursue a different transport agenda from elected leaders in local councils? There are many questions to be raised. The overall outcome though is a lack of progress in terms of increased cycling modal split, or simply a forecast saying what the future holds, which may make policy makers and planners lose interest in cycling-friendly cities' initiatives. One possible consequence of a lack of progress towards increasing cycling modal split may be the withdrawal of resources for developing tools measuring the cycle-friendliness of street networks. It would indeed be sad if such a withdrawal from cycling-friendly initiatives was as a result of transport planning models and other tools being short of data that can visualize 'high cycling modal share' policy options, and hence depriving policy makers of real choices.

Perceptions of the value of cycle planning

A second potential threat to mainstream use of GPS data is that cycle planning generally is considered a low-cost activity within the field of transport planning. For comparison, Keolis, a main public transport operator in Sweden, has 6300 full-time staff providing around 265 million public transport journeys per year (Keolis, 2013). This is equivalent to one member of staff per 42 000 public transport journeys, a figure that excludes people maintaining bus lanes and roads as well as people working at local authorities tendering bus services. As a basic comparison, a study of Swedish transport and urban planners identified that eleven planners in the sample spent 41 per cent or more of their office hours on activities related to cycle planning (Envall, 2009, p. 10). Four of these eleven people worked more than six out of ten office hours on cycle planning or promotion. The study had a 45 per cent response rate and the sample included the majority of the key cycling officers at Swedish local and regional authorities. Approximately 465 million cycle journeys per year take place in Sweden (SIKA, 2007). The number of full-time or near full-time staff working with cycle planning at Swedish local and regional authorities is therefore probably not much more than 20–25 people.[1] This corresponds to one less than full-time planner per 23 million cycle journeys, which is disproportionally lower compared to the public transport journeys ratio as previously mentioned.

Any transport planning officer who wants to contribute to a step change towards a more cycle-friendly city ought therefore not to neglect cost-effective data collection methods. This is because contemporary ICT for transport developments cutting costs for collecting and analysing cycling-related data would be much more likely to be widely adopted. Standardized ways of collecting cycling data would in addition be very beneficial regarding resource efficiency, because of the large number of people involved in cycle planning who can spend a small proportion of their working time on this activity.

1 The study had a 45 per cent response rate, so a rough estimate is that the total number of near full-time staff is 11/0.45 = 24 people.

3 THE CYCITY APPROACH TO ICT AND USE OF GPS DATA

3.1 Forming a Vision for the Use of ICT

A number of factors contributed to forming the CyCity approach which incorporates the use of ICT in cycle planning. These factors included:

- the weaknesses in traditional four-stage forecast tools in providing useful information on cycle-friendliness and accessibility by bicycle (Envall, 2007);
- the previously mentioned results from the baseline survey on the needs of transport planners (Envall, 2009);
- the mismatch between national and local levels in terms of perceptions about the future of cycling; as well as
- threats that the sustained and relatively strong interest in cycling-friendly urban policies will fade away without making a mark on planning practices for cycling.

Figure 8.2 illustrates the relationship between the four key factors underpinning the CyCity vision of making use of ICT in cycle planning.

All the above-mentioned perspectives helped form a long-term vision within the CyCity project.[2] This vision was expressed as employing ICT in the form of GPS-data from cycle users in order to better explain why cycling levels differ so widely between different urban areas in Europe, and from this, to provide better evidence for cost-effective policies to increase cycling and improve the evidence-base for cycle planning generally. This vision was then broken down into a goal to create a service package called BikeRoute; see Figure 8.2. BikeRoute is based on GPS data and geographical data on street networks.

Please note that the project simply rejected the perception of cycle planning as low value. However, the development of the BikeRoute service package has always been carried out knowing that funding often will be scarce when scaling up the use of ICT in cycle planning. Please also note that decision makers may need to be shown clear examples that collecting

[2] CyCity is a Swedish research programme established to improve knowledge about cycle planning processes. The programme also aims to increase understanding of the needs of cycle users. CyCity is financed by Vinnova, the Swedish Governmental Agency for Innovation Systems, and will run from 2010 to 2014. Its focus, volume and interdisciplinary nature makes it Sweden's most comprehensive research programme on cycle planning so far. A more complete presentation of CyCity can be found on the project website: www.cycity.se.

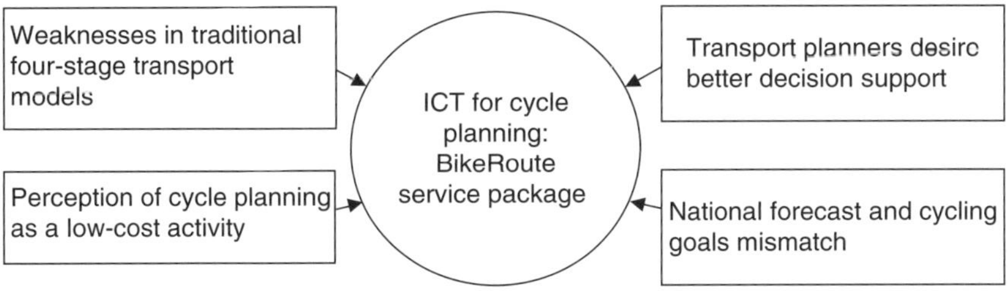

Figure 8.2 Illustration of factors forming a vision for the use of ICT within cycle planning, a vision leading to the development of the BikeRoute service package for cycle planning

GPS data on cycling behaviour is worth the effort, before spending the resources needed for improved decision support in cycle planning.

3.2 Explaining the Need for BikeRoute

The idea is that BikeRoute will help planners and elected leaders to form cost-efficient policies for increasing the cycling modal split or make cycle users more satisfied with their journeys. An important starting point for the service is to provide information on the quality of urban road networks from a cycling point of view, that is, 'How cycle-friendly is my city?'.

3.3 Some Previous Attempts at Measuring Quality

It is not easy to measure how cycle-friendly a city is. This is a key challenge highlighting the need to improve software for transport planning to explicitly include factors that are important to cycle users. Do cycle users have to stop at traffic lights and press a button and wait in order to get safely across, or are traffic lights sophisticated enough to detect cyclists approaching? Do people choosing to commute by bike have rights of way on local streets or do they repeatedly have to stop or take detours, making short car trips faster than cycling? Are there dedicated cycle facilities in the right place? A range of factors have varying importance for different user groups such as novice cyclists compared to experienced commuters.

The Netherlands is one of the world's most successful cycling countries, with many knowledgeable cycle planners. That is why when the CyCity partners describe the quality of transport infrastructure in cities, the Dutch experience is used as a reference point (Envall and Koucky, 2013).

The Dutch design manual for cycle traffic (CROW, 2007) presents five quality requirements:

- safety;
- continuity;
- directness;
- comfort;
- attractiveness.

These quality requirements have also been copied by several other countries to form their own guidelines for cycle planning. One such example is Ireland (National Transport Authority, 2011). But the Dutch quality requirements, as stated, are quite comprehensive. Not even the Dutch can always explain in more detail how, for example, continuity is measured or described. For example, everyone understands that many short sections of cycle paths which are not connected make it more irritating and difficult to ride, and maybe even less safe. But the Dutch design manual does not explain how to measure and concretely describe the qualities that are important for cycle traffic, at least not in a way that can easily be integrated in transport planning software. Thus, there is a need to explore options currently offered through ICT for transport to collect relevant data and support cycling planning based on such manuals and frameworks.

4 DEVELOPING THE NEW SERVICE PACKAGE

4.1 Evaluating Different GPS Devices

A CyCity sub-study evaluated a small number of different types of GPS device (Envall, 2012). The purpose of the evaluation was two-fold: first, to analyse different types of GPS devices' ability to accurately record cycle trips in an urban environment; second, to examine user-friendliness of different GPS devices when collecting data from cycle users.

CyCity compared different GPS devices in a simple way. Four different types of device were attached to the same bicycle. The devices included a Garmin GPS watch, a Holux GPS, a handheld Garmin GPS and a GlobalSat kit. The bicycle then travelled around denser as well as more modernistic (less dense) parts of Stockholm, Sweden. This provided the project team with several GPS tracks, one for each device. Tracks could then be compared to the actual route taken, comparing the accuracy and start-up times of different devices. Second, brief guidelines and a questionnaire were used to enable volunteer office workers in central Stockholm to evaluate some of the GPS devices. The GPS devices that performed better regarding start-up time and accuracy were used for this second step of the

Figure 8.3 CyCity GPS device used for data collection

evaluation. The two evaluations led to the selection of the most simple GPS device for data collection, a device with a simple On/Off button and an indicator (Figure 8.3).

4.2 Collecting Data Using New Technology

The CyCity project has so far used the selected GPS devices for data collection in two cities, Ljubljana, the capital of Slovenia, and the city of Linköping in Sweden. Over 800 cycle journeys were recorded by cycle users providing trip purposes and socio-economic data.

Volunteers cycling their typical everyday routes were recruited at work places in both cities. In total, around two hundred people recorded their cycle routes as well as data about their satisfaction with cycle routes or trip purposes. Hence, on average each individual recorded four one-way cycle trips, typically to and from their most common cycling destinations, for example from home to their workplace. Local partners in the two cities assisted with data collection.

4.3 Purpose of Data Collection

The main purposes of the data collection were to:

- Achieve a better understanding of cycling behaviour including increased knowledge of how people choose between alternative cycle routes and how cycling speed may differ on different types of route facilities.

- Definition of objectives & indicators ←
- Problem identification ←
- Choice of policy instruments
- Appraisal and prioritization ←
- Implementation
- Monitoring and evaluation ←

Note: Arrows indicate the main steps where the study and GPS data are believed to improve decision support.

Source: Adapted from May et al. (2003).

Figure 8.4 Steps in the transport planning process

- Increase knowledge of GPS devices' ability to provide reasonably accurate data when used for tracking cycling routes.
- Analyse the GPS data through a GIS database in order to describe the quality of the cycle network in a city. Hence the link with ICT takes here the form of GPS recording devices feeding data into GIS computer software.
- Make intelligible use of cycling GPS data in urban transport planning.

Perhaps most importantly, as illustrated in Figure 8.4, the data collection methodology and ideas were developed in order to address four key stages of the commonly formulated transport planning process (May et al., 2003). It was believed it would be useful to gather GPS data in order to measure the quality of cycle networks (definition of indicators) as well as for problem identification (what is good enough quality for cycling in our city?). The study approach when it comes to developing key indicators (see section 5.1 below) was adapted from a previous study on walkability and route choices (Envall, 2007, Appendix 6). A corner stone of the methodology is to consider any environmental attributes of streets on the basis of their cost and likely importance for cycling behaviour (propensity, route choice). Limited resources available for any real life transport planning study means that the total number of environmental attributes has to be restricted. Hence the development of indicators and visualizations were steered by the trade-off between cost and relevance for each environmental attribute (that is, type of cycle facility).

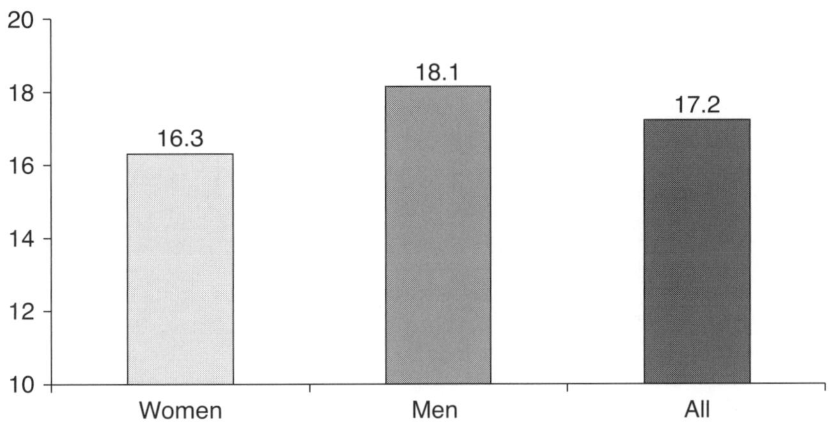

Figure 8.5 Average travel speed in the Linköping sample is 17.2 km/h

4.4 Survey Findings

This section highlights selected findings from the GPS study in Linköping, Sweden as well some results from a similar CyCity study in Ljubljana, Slovenia. Data for Linköping was collected in Spring 2011 with the assistance of Linköping City Council. Data for Ljubljana was collected in Summer 2012 with the assistance of the Urban Planning Institute of the Republic of Slovenia (UIRS).

Average travel speed

The average travel speed for the cyclists in the Linköping sample is 17 km/h. Men cycle somewhat faster than women (Figure 8.5). Travel speed is measured from start of movement when leaving the origin (for example cycle parking facility) to end at the destination (far end cycle parking).

Speed on different types of infrastructure

As shown in Figure 8.6, cyclists' average travel speeds differ on different types of infrastructure in Linköping. Average travel speed is observed to be the lowest on segregated cycle facilities. One reason for this is that segregated facilities in Linköping have separate traffic signals at junctions with main streets for motor vehicles and these signals have relatively short green light time windows for cyclists, causing delays (Engelson and Envall, 2013). Another possible reason for slower travel speed is that segregated paths in Linköping are often shared between pedestrians and cyclists.

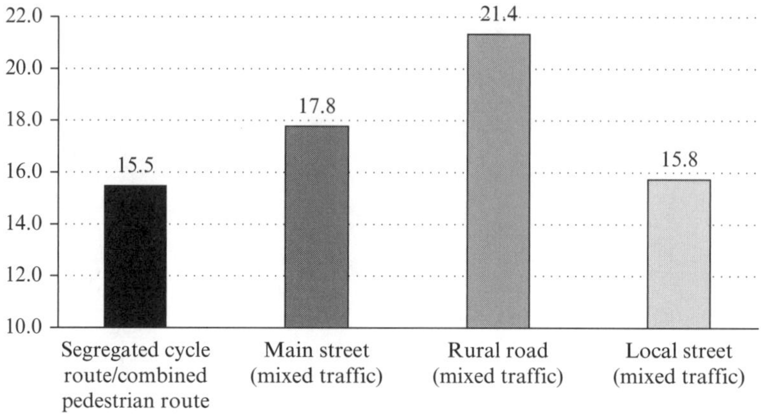

Figure 8.6 Differences in cycle speed (km/h) on various types of cycle infrastructure

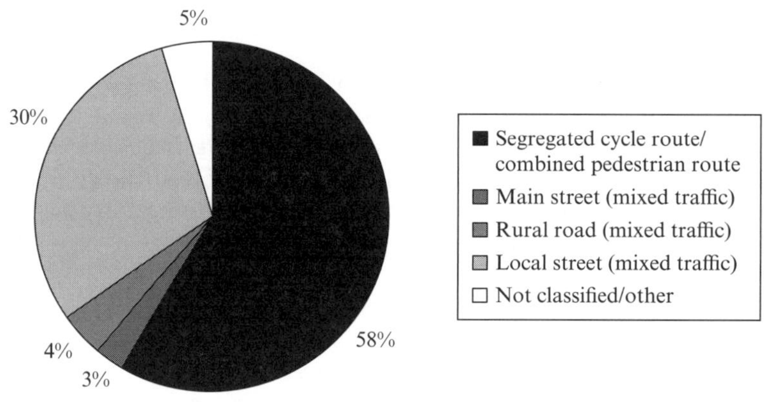

Figure 8.7 Proportion of distance travelled on different types of infrastructure in Linköping

Data about travel speed is valid for the warmer period of the year, that is, not winter.

Travel distance on different types of infrastructure

Figure 8.7 illustrates that around one third of the kilometres cycled in the sample from Linköping are carried out in mixed traffic conditions, that is, on roads and streets with no dedicated cycle facilities where cars and bicycles share the same lanes.

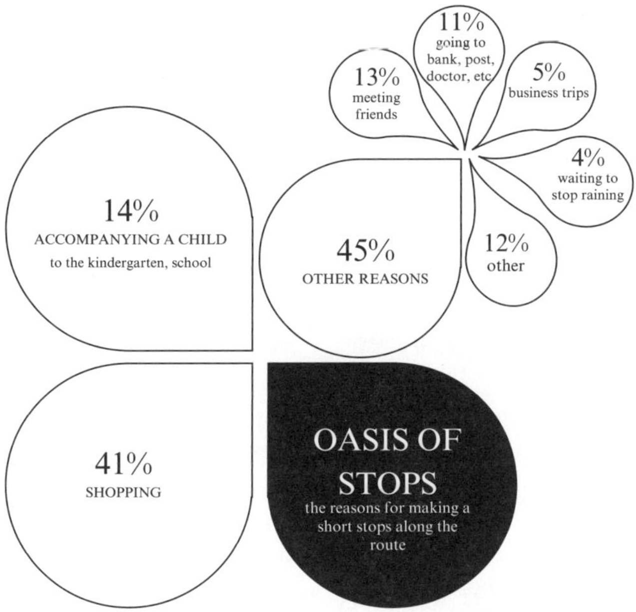

Source: Illustration by UIRS

Figure 8.8 Reasons for short stops along the cycle route in Ljubljana

Stop frequency for carrying out errands

Around one in three trips in the data sample from Ljubljana (30 per cent) included a short stop carrying out an errand. The same figure for Linköping was lower. In Linköping around one in five trips (22 per cent) included a short stop. The most common reason for carrying out a short stop en route in Ljubljana was shopping; see Figure 8.8.

5 RESULTS

5.1 Software Features

So far, the work has resulted in the following innovations:

- a service package and software for automated analysis of GPS route data;
- creation of maps and graphs to visualize and compare cycling quality of infrastructure;

- scripts that can be used to evaluate the impact of new cycling infrastructure, for example the ability of a new segregated cycle route to increase the continuity of the cycling network.

Based on the analysis of the data and previously collected evidence, the project identified a small number of key indicators for a realistic but comprehensive understanding of the cycle-friendliness of a particular city's street network. These indicators are realistic in terms of evaluation costs when making use of GPS data. The three key indicators which are reviewed in this section are:

- travel speed;
- route continuity;
- access to segregated facilities.

The purpose of key figures drawn by BikeRoute is to present important aspects of the cycle infrastructures quality, that is, how easy and comfortable it is to ride a bicycle. Utilizing this data, it is also possible to evaluate overall goals set for traffic and transport planning. City-level strategic transport plans may include targets such as to measure 'cycling speed' (City of Copenhagen, 2013, p. 7) or targets that the 'accessibility to cycle facilities will increase'. It may be argued that such relatively vague policy objectives may be less effective in bringing about a step change towards more cycling-friendly cities if not defined and measured quantitatively. In any case, the ICT approach developed within CyCity should make it possible to evaluate cycling measures *ex post* and therefore learn more about different measures' impact on travel speed. A key benefit is that by using GPS devices new knowledge can be gained at a significantly lower cost than if an on-site survey team collected the data.

5.2 Travel Speed

Travel speed is a major factor for cyclists. Are there any bottlenecks that force cyclists to make frequent stops? Surveys from, for example, Copenhagen (Köpenhamn, 2012) show that the main motivation to start cycling is that the bicycle is faster than alternative modes of transport.

BikeRoute analyses the travel speed of the cycle network with a high level of detail through GPS data generated by actual cyclists. In this way, the city gets a concrete baseline picture of cyclists' accessibility, route preferences and bottleneck locations; see Figure 8.9 for an illustration. Travel speed measurement can also demonstrate whether accessibility levels are different for different types of cycling infrastructure, various cycle groups as well as the type of infrastructure cyclists use the most. Figure 8.10

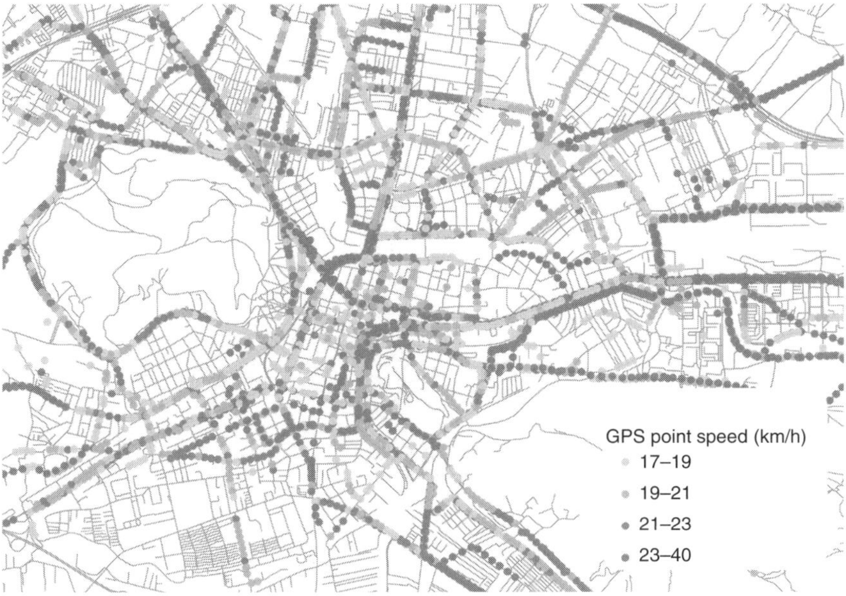

Source: Image: CyCity

Figure 8.9 *Differences in actual cycle speeds within the cycle network in Ljubljana, Slovenia*

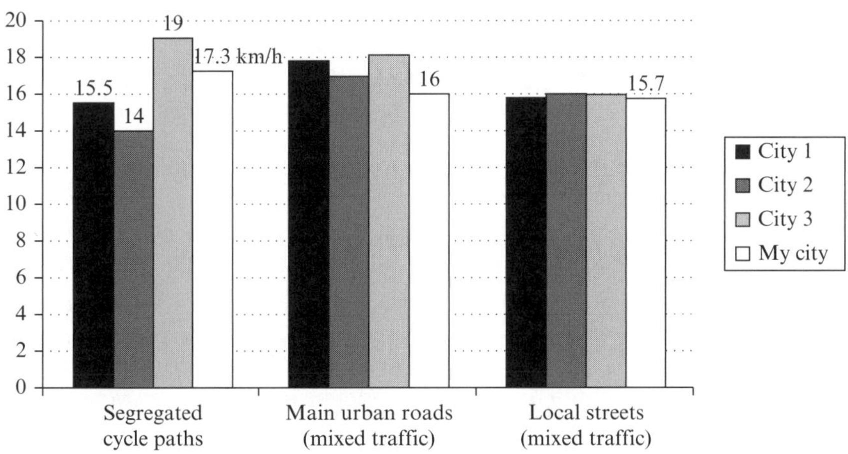

Source: Image: CyCity.

Figure 8.10 *Comparing differences in real cycle speed on various types of cycle infrastructure in different cities*

Note: Light grey shows the connected network where black and dark grey shades show small isolated islands.

Figure 8.11 The continuity of the cycle infrastructure in Linköping expressed by the number of 'islands' in different shades

illustrates how speed data from different cities can be benchmarked for increased understanding of user problems and opportunities.

5.3 Continuity

The interconnection of cycle infrastructure is important for cyclists, as reported in the Netherlands (CROW, 2007). Unfortunately, many cycle paths end abruptly and cyclists are forced to ride in mixed traffic to continue their journey or to enter the next section of the cycle path. This can be frustrating for several cyclists, but can also be a deterring factor leading to opting out of cycling completely, particularly for novice cyclists and children. A short missing section may make a big inadvertent impact. BikeRoute helps to visualize and measure how continuous the cycle infrastructure of a city is and how effective an increase of the network continuity could be. Figure 8.11 provides an illustration that can be requested when carrying out a strategic transport analysis.

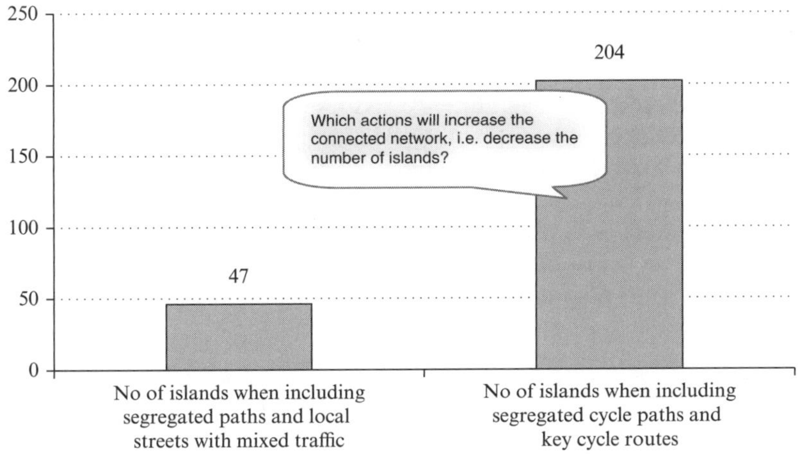

Source: Image: CyCity.

Figure 8.12 Number of 'islands' using two different network measures

As shown in Figure 8.12, the continuity analysis includes two different indicators at a more detailed level. An 'island' in the analysis is defined as a cycle facility or connected group of cycle facilities that are linked only to other cycle facilities by making use of a road with mixed traffic (with cars, HGVs) above a certain speed limit (>30 km/h). The left-hand column in Figure 8.12 assumes that experienced adult cyclists are quite happy with cycling on mixed traffic streets as long as car speed and motor vehicle numbers are relatively low. The right-hand column only deems segregated facilities as 'good enough', appropriate for the user needs of novice cyclists and children.

5.4 Access to Segregated Cycle Facilities

The third indicator is designed to give a quick understanding of the potential benefits of providing new cycle facilities in different areas. BikeRoute matches population data with the available cycle infrastructure in order to clearly visualize the current situation in the city. The availability of cycle infrastructure is expressed in a number of key ratios, such as the proportion of the day and night population that are within the catchment area for segregated or other types of cycle facilities, for example within 200m and 400m. Data is presented on maps and graphs.[3]

---

[3] For more information see: http://cycity.se/eng/.

6 BARRIERS FOR FULL-SCALE IMPLEMENTATION

To sum up, there are at least three key issues for full-scale implementation of the new decision in support of cycle planning based on GPS data and analysis of GIS street networks. These are summarized as:

- The flawed perception that currently used transport planning software and transport planning approaches deal with cycling travel speed, continuity and accessibility in a better way than they actually do.
- The need for standardized and accurate digital GIS data about cycle infrastructure and street networks, in contrast to the fact that many transport planners are not experienced GIS users. This is one of the areas where ICT for transport can clearly make a positive contribution.
- The perception that improved decision support for cycle planning is not really needed and that GPS data on cycling is perceived to offer low value for money.

Previous research has described how transport planning models and software are used in practice. Examples of such studies are Atkins (1987) and Tennøy (2003, 2004). Tennøy (2003) found, for example, that one third of the transport planning documents investigated did not mention uncertainties of forecasts presented. Models are furthermore presented as 'multi-modal' despite the fact that they do not include walking and cycling route quality (IHT, 1996, p. 15; MVA, 2002). Models that are seen to be multi-modal, despite not being the case, are obviously a barrier for full-scale implementation of approaches making use of GPS data for cycle planning. A planner who thinks that the model study that someone else has carried out fully includes cycling would of course be likely to show little interest in embarking on collecting GPS data or carrying out specific cycling modelling or assessments. Notwithstanding, there seems to be an increased awareness of the limitations of traditional software for transport planning when it comes to how these do or do not fully analyse cycling. That would then leave two remaining barriers, namely the need for standardized and accurate street network data and the perception that there is little to gain from more sophisticated approaches to cycle planning, as discussed in section 2.2.

7 CONCLUSIONS

ICT in the form of cyclists' GPS data is a powerful source for analysing the quality of urban transport networks. Cyclists' GPS data gives planners a better understanding of cycling behaviour, route choices and travel speed. The use of GPS devices also has the potential to cut costs for data collection and quality analysis compared to traditional manual data collection methods. GPS data on cycling is furthermore one step en route to giving decision makers more well-balanced decision support, explicitly illustrating the quality of urban transport networks from a cycling point of view. This data should help implement walking and cycling-centric planning perspectives in a cost-effective and evidence-based way.

Generally the quality of GPS data on routes cycled was deemed good or fair in the two cities where GPS data was collected. However, there are potentially some issues with data quality in dense urban areas. Tall buildings and narrow streets lined by building façades make cyclists' GPS recordings less reliable (Envall, 2012). This needs to be considered when choosing an appropriate GPS device for recording cycle routes. The fact that GPS devices are not completely accurate also makes specialist software necessary for making intelligible use of the data.

The study approach illustrated in this chapter can be applied in other cities. This is relevant because many cities are experiencing an increase in cycle trip numbers for commuting purposes, for example London. Even cities not offering people safe and convenient cycling infrastructure are experiencing a rise in cycling flows, although from relatively low levels. A remaining question is, what would cycling numbers be in these cities if they had superior quality transport networks from a cycling point of view? There is also potential in the widespread use of smartphones for collecting cycling GPS data. This is currently being explored by the CyCity project and without doubt by many others.

ACKNOWLEDGEMENTS

This work is a part of the Swedish CyCity Research Programme (www.cycity.se). Project partners are TUB Trafikutredningsbyrån AB, WSP Sweden AB, Koucky & Partners AB and VTI. Linköping City Council and the Urban Planning Institute of the Republic of Slovenia (UIRS) have also contributed to the project. CyCity is financed by Vinnova, the Swedish Governmental Agency for Innovation Systems.

REFERENCES

Atkins, S.T. (1987) The crisis for transportation planning modelling. *Transport Reviews*, 7(4), 307–25.

BBC (2013) Bicycles outsell cars in most EU countries in 2012. BBC News Europe, 3 December. BBC, London. Available at: http://www.bbc.co.uk/news/world-europe-25209551.

Brundell-Freij, K. and Envall, P. (2012) Cykel och ITS: Förstudie med förslag till handlingsplan för Storstadsforums arbete. Stockholm: WSP Sverige AB.

City of Copenhagen (2013) Copenhagen City of Cyclists: Bicycle Account 2012. Technical and Environmental Administration Traffic Department, City of Copenhagen. Available at: http://subsite.kk.dk/sitecore/content/Subsites/CityOfCopenhagen/SubsiteFrontpage/LivingInCopenhagen/CityAndTraffic/~/media/4ADB5281 0C484064B5085F2A900CB8FB.ashx.

CROW (2007) Design manual for bicycle traffic. Ede, the Netherlands: CROW.

DfT (2012) Road transport forecasts 2011: Results from the Department for Transport's National transport model. January. London: Department for Transport. Available at: https://www.gov.uk/government/uploads/system/uploads/attachment_data/file/4243/road-transport-forecasts-2011-results.pdf.

EC (2007) Sustainable urban transport plans: Preparatory document in relation to the follow-up of the thematic strategy on the urban environment. Main document 25 September. Technical Report – 2007/018. Luxembourg: Office for Official Publications of the European Communities. Available at: http://ec.europa.eu/environment/urban/pdf/transport/2007_sutp_prepdoc.pdf.

Emanuel, M. (2012) *Trafikslag på undantag: Cykeltrafiken i Stockholm 1930–1980*. Stockholm: Stockholmia Förlag.

Engelson, L. and Envall, P. (2013) *Cyklisternas fördröjningar i korsningar. En undersökning baserad på GPS-data om cykelresor i Linköping*. December. CyCity report. Stockholm: WSP Sverige AB. Available at: http://www.cycity.se/docs/2013-12-09_Cyklisters_fordrojningar_slutlig.pdf.

Envall, P. (2007) Accessibility planning: a chimera? PhD Thesis, Institute for Transport Studies, University of Leeds, Leeds. Available at: http://www.trafikutredningsbyran.se/files/Envall_PhD.pdf.

Envall, P. (2009) Decision support for cycle planning: results from a questionnaire survey in Sweden. TransportMistra. Institute for Transport Studies, University of Leeds, Leeds. Available at: http://www.cykelplanera.se/Survey.pdf.

Envall, P. (2012) Användning av GPS-mottagare för cykelstudier: Lämplighet för ruttvalsanalys mm. 23 March. CyCity working paper available from the author.

Envall, P. and Koucky, M. (2013) Varför behöver man beskriva trafiknätets cykelvänlighet? CyCity report, August.

Floater, G., Rode, P., Zenghelis, D., Motero-Carrero, M., Smith, D., Baker, K. and Heeckt, C. (2013) Stockholm – Green Economy Leader report, London: LSE Cities.

IHT (1996) Guidelines for developing urban transport strategies. London: The Institution of Highways & Transportation.

Keolis (2013) Information on Keolis' Swedish website. Accessed October 2013 at www.keolis.se.

Köpenhamn (2012) Cycle Super Highways. Copenhagen: Teknik och Miljöförvaltningen.

Lundin, P. (2008) Bilsamhället: ideologi, expertis och regelskapande i efterkrigstidens Sverige. PhD Thesis. Stockholm: Stockholmia förlag.
Marsden, G., Frick, K.T., May A.D. and Deakin, E. (2010) How do cities approach policy innovation and policy learning? A study of 30 policies in Northern Europe and North America. *Transport Policy*, 18(3), 501–12.
May, A.D. et al. (2003) *A Decision Makers' Guidebook*. Prospects Deliverable No. 15. Institute for Transport Studies, University of Leeds, Leeds. Available at: http://www.ivv.tuwien.ac.at/fileadmin/mediapool-verkehrsplanung/Diverse/Forschung/International/PROSPECTS/pr_del15dmgen.pdf.
MVA (2002) South and West Yorkshire multi modal study: final report. Manchester: MVA Ltd.
National Transport Authority (2011) National cycle manual. Dublin: Irish National Transport Authority. Available at: http://www.cyclemanual.ie/.
NCE (2014) Better growth, better climate: the new climate economy report. Global report. Washington, DC: Global Commission on the Economy and Climate.
Niska, A. et al. (2010) Metoder för skattning av gång- och cykeltrafik: Kartläggning och kvalitetsbedömning. VTI report 686. VTI, Linköping.
SIKA (2007) Resvaneundersökningen 2005–2006. Swedish National Travel Survey. Östersund: Statens Institut för Kommunikationsanalys.
Sloman, L., Cavill, N., Cope, A., Muller, L. and Kennedy, A. (2009) Analysis and synthesis of evidence on the effects of investment in six cycling demonstration towns. Report for Department for Transport and Cycling England.
Tennøy, A. (2003) Prediksjoner og usikkerhet trafikfaglige rapporter i KU. NIBR-report 2003:13. Oslo: Norsk Institutt for by- og Regionforskning [in Norwegian].
Tennøy, A. (2004) Transportanalyser i planleggningen: til hinder for baerekraftig byutveckling? Notat 2004:121. Oslo: Norsk Institutt for by- og Regionforskning [in Norwegian].
Trafikverket (2012) Persontransporter: Underlagsrapport. Publication no: 2012:121. 27 April. Borlänge: Trafikverket. Available at: http://publikationswebbutik.vv.se/upload/6731/2012_121_Persontransporter.pdf.
WSP (2012) Framkomlighetsstrategi: metodutveckling för parkeringsutredning på huvudgator. Project no. 1014 4992. Stockholm: WSP Sweden AB.

PART III

Policies promoting ICT for transport: contemporary and future challenges

9. Reversing the order: towards a philosophically informed debate on ICT for transport

Heidi Herzogenrath-Amelung, Pinelopi Troullinou and Nikolas Thomopoulos

1 INTRODUCTION

Any microprocessor-based activity utilizes data which can be saved and increasingly correlated. This trend, which took off during the first decade of the twenty-first century, gave birth to the notion of Big Data. The enormous number of applications of Big Data are still to be fully realized, but the enhanced precision of location based data and access by third parties – that is, parties other than the mobile telecommunication network operators and the mobile device user – may raise concerns over the use of such data. Driven by growing global concerns for legitimizing innovations prior to their diffusion, this chapter aims at minimizing the unintended social and technological consequences of ICT development without any prior assessment of the potential impacts (Owen et al., 2009). Big Data has the potential to unlock promising opportunities, but at the same time it introduces new challenges for the fields of spatial economics and transport (Dekkers and Rietveld, 2009). ICTs are increasingly used in transport contexts to make travel more efficient, cost-effective and also convenient and entertaining through gamification. At the same time, such technologies enable increasingly comprehensive surveillance from the data gathered via devices and infrastructure. Thus, this chapter is a theoretical contribution to the debate concerning the balance between the positive contributions ICTs can make in the transport sector and the risks arising from the gathering of increasing amounts of personal data. In order to address the interdisciplinary nature of this debate, this chapter draws on the following theoretical approaches: medium theory,[1]

[1] Medium theory is distinguishable from media and communications theory more generally by its emphasis on the impact of particular media and communications technologies, rather than on the content they communicate (Meyrowitz, 2008).

philosophy of technology, critical theory and surveillance studies. It foregrounds the dual use of ICTs in transport contexts offering policy-relevant recommendations to inform the inclusion of ethical concerns in the design stage of ICTs for transport.

The chapter begins by emphasizing the ways in which transport is becoming increasingly reliant on ICT applications as part of the growing dependence of modern life on digital technologies more generally. We then examine the benefits arising from an increased use of ICTs for transport in response to the challenges posed by the need for greater mobility and flexibility that defines contemporary life and work patterns. We draw attention to the discourses emphasizing these positive contributions such as convenience, cost-effectiveness, reduction of environmental pollution and healthier lifestyles. The Oyster card and the Chromaroma game are discussed as examples illustrating our core argument that presenting these applications purely in terms of their potential to improve individual and collective mobility misses some of the more problematic aspects arising from location-based services (LBS) and their modes of data collection (Bagchi and White, 2005). Big Data are not discussed in detail, but are referred to as another upcoming challenge in the last section. We then map what we consider to be the key ethical, social and legal implications, arguing that ICTs for transport need to be considered within the wider context of surveillance and the exploitation of digital labour (Andrejevic, 2011; Fuchs, 2011). The chapter concludes by suggesting that a technophobic approach should not be adopted, but instead that philosophy of technology could be valuable in informing an *ex-ante* framework for policy-makers, software developers and researchers. Such a framework would be useful in shaping the development of ICT for transport in advance of any legislative measures which merely mitigate risks and limit potential misuse *ex post*.

2 CONTEMPORARY APPLICATIONS OF ICT FOR TRANSPORT

A wide variety of ICTs are being used in contemporary transport infrastructure and a lot of these aim at promoting better use of economic, social and environmental resources. For example the Oyster card constitutes an integrated ICT for transport scheme which can be used on all public transport modes across London, for example, underground, buses, Docklands Light Railway, river boats. From an operator's point of view this is a hugely beneficial development since it allows the analysis of travel behaviour data and management of transport demand in ways that were not possible in the twentieth century. The increased use of Variable Message Signs (VMS) on motorways or public transport interchanges is another example of positive applications of ICT for transport informing travellers whilst allowing transport operators to direct traffic according to network capacity. The latter has noteworthy implications both for research and policy, particularly in the light of contemporary efforts to increase public transport demand and lower carbon emissions in densely populated urban areas (Nakamura and Hayashi, 2012; TfL, 2012). In conjunction with other transport measures facilitated through ICT, such as congestion charging schemes using Automatic Number Plate Recognition (ANPR), these developments contribute to sustainable urban transport at various degrees (Givoni, 2012). At the same time, the introduction of Wi-Fi and other ICT infrastructure in public transport has ameliorated passengers' experience and the use of travel time as highlighted in the literature (Lyons and Urry, 2005; Mokhtarian and Salomon, 2001; Mott MacDonald, 2009).

In addition, a plethora of transport-related smartphone apps have been developed during the last decade,[2] all of which demand access to a range of user information in exchange for certain benefits such as convenience, safety, eco-friendliness or travel cost reduction. Although it is not among the aims of this chapter to offer a comprehensive overview of such applications, a brief overview of applications for different transport modes is offered to illustrate user benefits.

The partnership between PleaseCycle[3] and Cyclescheme in the UK is a contemporary example of the potential benefits of ICTs for transport for cyclists through gamification, whilst addressing congestion and emissions reduction. Cyclists only need to register their bicycle trips in order to gain access to the scheme rewards. In this case employees can benefit through retail outlet vouchers, employers can benefit by demonstrating that they are reducing their carbon footprint and urban areas can benefit through reduced car traffic and lower emissions in the local atmosphere. Similarly, users are able to meet individual targets, for example to walk more and improve their personal health and well-being while receiving rewards. SMART[4] is an application offering an additional feature, a heatmap based on crowdsourced information so that users in the Netherlands can

---

[2] Ideas in Transit offer a snapshot of such applications until 2012 (Ideas in Transit, 2012).

[3] PleaseCycle: http://www.pleasecycle.com. The use of brand names and trademarks in this chapter aims only at illustrating the arguments made by the authors. No copyright or ownership claims are implied through the use of these brand names and trademarks, which belong to their respective registered licence holders.

[4] SMART: https://play.google.com/store/apps/details?id=nl.enschede.smart&hl=en.

avoid congested routes. Another application which utilizes crowdsourced information to offer real-time traffic updates reducing travel time and fuel costs is Waze,[5] which is an application widely used across the US due to the high car use by commuters. Such applications may be enhanced in the future through vehicle-to-vehicle (V2V) communication offering real-time traffic updates as discussed in Chapter 7. Along the same lines, Taxibeat[6] is an application automatically detecting a user's location, which can call a taxi via a single tap on one's smartphone. The added value in this case is that the application allows the selection of a taxi (for example, vehicle, driver) based on previous evaluations, which can enhance the user's experience. Moreover, new opportunities appear for the vehicle insurance sector since mobility tracking allows the charging of insurance premiums based on the actual distance travelled. Users may also benefit from the increased application of ICTs through services such as Friendsurance,[7] which aims at lowering the insurance cost for each individual user. Moreover, safety concerns over vulnerable social groups, namely children, the disabled or the elderly may also be addressed through ICTs for transport by enabling carers to monitor their trips and locations in time.

What should be evident from the aforementioned is that personalized travel planning, rewards, real-time updates about traffic, public transport or parking availability, e-tolls, lower vehicle insurance premiums and increased safety are all benefits realized through ICT for transport applications. It is thus apparent that various benefits for users, manufacturers, local authorities and traffic managers derive from such use of ICTs within the transport sector (Bekhor et al., 2013; Rose, 2006; Steenbruggen et al., 2013; Thakuriah and Geers, 2013). On the other hand these applications require direct and constant access to a user's:

- current location;
- home address;
- transport mode;
- mobile device details;
- software details;
- payment details;
- hobbies and interests.

The argument in this chapter is that despite the apparent benefits for research and policy making, there are important issues surrounding such

---

5 Waze: http://www.waze.com.

6 Taxibeat: http://www.taxibeat.com.

7 Friendsurance: http://www.friendsurance.com.

user data and users' awareness of this data collection (FTC, 2012). This may raise concerns for children (UNCRC, 1989) or other vulnerable social groups as their privacy is important. A debate about safety or privacy priority (Eriksson and Bjornskau, 2012) arises then, which is further discussed in this chapter. Additionally, there are wider implications of this issue for any mobile tracking device user since smartphone apps such as PleaseCycle may encourage employer surveillance over their employees in certain industries, for example logistics (Kanngieser, 2013).

A common criticism of such privacy concerns has been the inability of individuals to make rational decisions in certain ICT scenarios due to bounded rationality (Acquisti and Grossklags, 2005). Yet Potoglou et al. (2013) have addressed the latter issue through a stated preference survey which corresponds with common practice in the transport sector. They conducted three stated choice experiments which confirmed the privacy paradox (Barnes, 2006), namely that individuals state that they are concerned about their privacy but they are not willing to pay to protect their personal data. Consequently, it remains to be seen how these technological transformations, including the highly anticipated benefits of the Galileo positioning service to be launched in Europe (see Chapter 3), will be managed in practice by users, corporations and policy-makers to address these issues.

3 ICTs AND THE DISCOURSES SURROUNDING TRANSPORT APPLICATIONS

In this section the focus moves from the benefits of the increased application of ICTs for transport to the implications of how these benefits are marketed to consumers. The emphasis on usability in marketing communications (for example convenience, ease of use) and the fact that these ICT applications seemingly emanate from self-contained gadgets (for example smartphones, tablets) encourage the user to focus only on their immediate benefits, disregarding any risk that ensues from these applications.

In the context of the move towards ever greater mobility, the application of ICTs in the transport sector for instance, promises to make planning and booking transactions and travel itself safer, as well as more cost- and time-efficient (Dijst, 2009). The 'Oyster' smartcard in London, for example, has replaced older paper-based tickets as well as the need for individual ticket systems for different kinds of public transport (underground, bus, light rail), making travel 'cheaper' and more 'flexible' (Transport for London n.d., a). Such integrated ticketing systems are often called 'smart' because they allow users to travel on different modes of transport or on modes provided by different transport operators (HMC, 2013). How ICTS are making travel 'smarter' is also the argument used in the marketing communication of smartphone applications such as the Dutch SMART app and the US traffic and navigation app Waze: the latter especially emphasizes the benefits of cost- and time-reduction ('saving everyone time and gas money') (Waze, 2013). Marketing communications foreground these positive uses because, as Murakami-Wood and Ball (2013) point out, any association with problematic practices such as the collection of personal data would have undesirable results for the market as consumers would not align their interests with those of the cooperation. Thus discourses such as 'convenience' and 'time-effectiveness' serve to maintain the alignment between the interests of the company and the consumers, preventing an alternative consideration of the consumer's technological activity or device.

The crucial question is how precisely this alignment takes place. We suggest that marketing discourses conceptually 'package' digital developments in a way that produces specific attitudes or beliefs amongst users, a process which can be understood from the perspective of Althusser's theory of interpellation. According to Althusser (1970), individuals are constructed as subjects by responding to and thus perpetuating the ruling ideology. Althusser uses the example of a policeman who shouts out to a member of the public, producing an instantaneous reaction in this person in so far as he or she turns around to await the officer's further instructions. This process of 'hailing', according to Althusser, is how ideology is maintained – through language. Or, as Fairclough has argued more recently: 'Ideologies are closely linked to language, because using language is the commonest form of social behaviour' (2001, p. 2). The individual's response reproduces the ideology, and thereby the power of the 'Ideological State Apparatuses' (ISAs) from which it originates – in this case the law, but the media, or the market, as Andrejevic (2013) argues, equally function as ISAs that determine our beliefs, actions and decisions.

Based on Althusser's model of the dissemination of ideology through language, it becomes possible to understand how marketing discourses shape our beliefs and attitudes regarding the use of ICT applications. The location-based app Google Maps, for instance, markets its service via the invitation to 'explore new' and 'share favourite places with friends', emphasizing the positive attributes of 'shared experiences' (Google Inc.). Surrounded by discourses of 'sharing' and 'exploring', the marketing of the applications seems to suggest to the user that they merely facilitate a variety of desirable experiences. Figure 9.1 builds on Althusser's theory of interpellation in an attempt to model the effects of the marketing discourses on consumer attitudes and actions. At the centre of the model

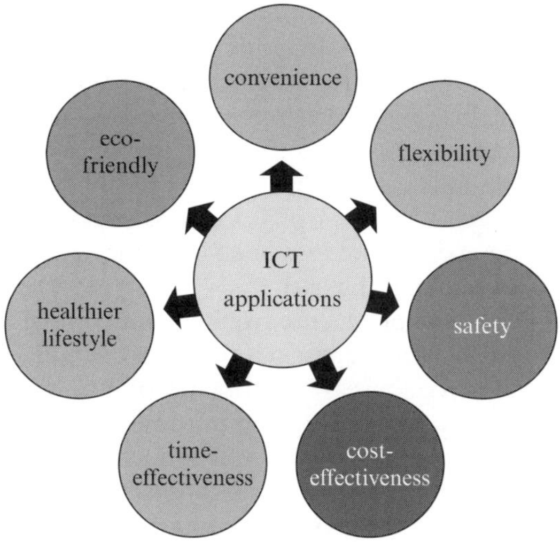

Figure 9.1 Discourses surrounding ICT applications

is the ICT application, where the surrounding bubbles contain some of the most common marketing rhetoric in the transport context, such as time-effectiveness, convenience and eco-friendliness. Similar to the hailed subject in Althusser's example, the user of ICT applications is hailed as an uncritical and enthusiastic consumer by marketing discourses. The desirable experiences he is promised act as a deflective layer that distract attention away from the technologies themselves and any potential risks, towards their prospective benefits.

4 DIGITAL CONVERGENCE AND EMERGING RISKS

Sections 2 and 3 explored how the discourses surrounding the implementation of ICT in transport contexts are successful in creating positive connotations of these technologies amongst users (Figure 9.1). Oyster smart card users, for instance, with a monthly travel card subscription are often not aware of the nature and extent of information held about them, but even for a visitor who purchases an Oyster card by debit or credit card this includes not only their full name, address, telephone number, email address and password, but also encrypted bank details. Crucially, this information is then supplemented by records of locations, dates and times of journeys made on transport systems where Oyster is in use. Each time the card is swiped, its user adds another piece of information to a user's digital profile, or data double (Haggerty and Ericson, 2000): the digital 'footprint' that each individual is leaving behind through his/her use of ICTs. For the user, however, travel constitutes a socio-cultural practice and as a result of how ICTs in the transport sector are marketed he/she sees these as a continuous improvement of this practice. Thus the question of risks emerging from the use of ICT for transport – for example what happens to the data accumulated via the use of the Oyster smartcard – is not immediately relevant for the user.

In the Oyster card example, its use results in millions of new data additions daily (TfL, 2013), which contributes to what is now called Big Data and consists of several databases. The principle of purpose limitation – the idea that data may only be used for the purposes for which it was originally collected – becomes meaningless as soon as attention is paid to the small print of the Oyster terms and conditions: Transport for London (TfL) has the right to 'share Oyster information with their subsidiaries and service providers' which also is comparable with other services used also for marketing purposes (Transport for London n.d., b). Chromaroma (2013) may be such an example, since it offers an overview of all the trips of an Oyster user in an attempt to introduce gamification to the everyday life of Londoners.[8] Similar data-sharing practices take place for smartphone applications such as PleaseCycle, Taxibeat and Waze. Further to this, following the 1998 Data Protection Act, TfL is obliged to comply with requests made by the police and other law enforcement bodies for disclosure of personal data. In 2006, *The Guardian* reported 243 requests by the Metropolitan Police for release of personal information of Oyster card holders; by 2010 this figure had risen to 6576 requests, with only 810 requests denied (Xynou, 2012). This means in over 85% of all cases the information asked for was released. In addition it should be noted that a range of other bodies have also made requests to TfL for disclosure of information held on Oyster card users, amongst others solicitors, the UK Border Agency and city councils (Xynou, 2012).

What the above example makes clear is that data collected via a transport 'smart application' are easily merged with data collected from other applications or sources in an attempt to create Big Data-bases. In this case information from an Oyster smartcard or Chromaroma is combined

8 Chromaroma is an entertaining game based on the data created by users and left behind unnoticed. Mudlark, the Oyster-based travel game developer, aims to help users make 'smarter' decisions about their lives based on their own travel data overview through this game. This is anticipated to be achieved by building communities of users whilst overcoming the fear of creating a dystopian database society (Chromaroma, 2013).

with data from the individual's credit or debit card. The merging of data from two seemingly self-contained units is enabled by the basic operating principle of digital information and communication technology, the processing of binary code (a sequence of 0s and 1s). Government agencies and private businesses alike are harnessing the benefits of this simple principle: the foundation of Big Data, the Open Data initiative in the UK and the National Data Warehouse for Traffic Information in the Netherlands are developing databases and combining them in ways not previously possible, which has expanded the range of their field of action. What presents itself to the user on the other hand is an interface that promises precisely the positive experiences listed in Figure 9.1: convenience, flexibility, and time-effectiveness. Users' control over their data is replaced with usability – as evidenced by Potoglou et al. (2013). Or, as the medium theorist Friedrich Kittler argues:

> [t]he general digitization of channels and information erases the differences between the individual media. Sound and image, voice and text are reduced to surface effects, known to consumers as interface. Inside the computers themselves everything becomes a number . . . With numbers, everything goes. Modulation, transformation, synchronization; delay, storage, transposition; scrambling, scanning, mapping – a total media link on a digital base (Kittler, 1999, p. 1)

The digitization of information processing allows for contact-less transfer of information between any number of devices and databases without the explicit consent, or in some cases even the knowledge of the user. Using smartphones introduces further risks since location data and photos can be easily stored and transmitted from mobile devices without a user's explicit consent. What emerges from these considerations is that ICTs for transport which include location, distance, duration, frequency and mode data, form part of and should be considered from the perspective of, a single informational system. Within this system the user's personal information circulates increasingly seamlessly as part of a great 'amorphous flow' of data that to the user is 'unknown and seemingly uncontrollable' (Zurawski, 2011, p. 510). The revelations of the scope of the surveillance programmes run by the American National Security Agency ('PRISM') and its British counterpart ('Mastering the Internet') in the summer of 2013 have shown in unprecedented clarity the extent to which the flow of data is beyond the control of the user as well as the increasing cooperation between states and commercial entities.

The 'total media link on a digital base' mentioned by Kittler has so far been seen from a largely celebratory 'digital convergence' perspective which allows the end user to traverse multiple media platforms in search of the total information and entertainment experience (Jenkins, 2006). But as the example of the Oyster card shows, with convergence there comes what Boyd (2008) refers to as the 'collapse of contexts': the user of an Oyster card originally provided his personal data in the context of transportation in the London area. The contextual boundaries that originally separated the spheres within which data was gathered or made available and assigned specific purposes to the data collected, are thus becoming increasingly porous. As Boyd points out, the 'collapse of contexts' that results from digital convergence has significant implications for the individual, as it results in a lack of control over his or her data, but even more significantly, it results in a need to renegotiate the terms of privacy.

Whenever surveillance has come under scrutiny, privacy has been one of the main concerns. Yet, as David Lyon argues, we are far from having reached an agreement on how to define privacy, even in the West, where one might assume a certain homogeneity of opinion exists about personal, physical privacy, for instance. In the West bathroom activities for instance are largely considered private, whereas Indian households lack such a concept (Lyon, 2007). Helen Nissenbaum's (2004) argument that privacy is 'contextual' (that is, what is private in one context might not be in the other) has been highly influential as it appears to solve issues of cultural difference. We would agree with Boyd, however, that 'information is not private because no one knows it; it is private because the knowing is limited and controlled' (Boyd, 2008, p. 18).

5 ETHICAL, SOCIAL AND LEGAL CONSIDERATIONS OF ICTs FOR TRANSPORT

Urban, geography and transport studies have emphasized the benefits of the growing application of ICTs in smart cities in general and in the transport sector in particular (Bagchi and White, 2005; Bannister, 2012; Cottrill et al., 2013; Midgley, 2009; Shaffer et al., 2012; Tyrinopoulos, 2011; Xiao et al, 2012), both in terms of personal user experience through reducing journey times, improving cost-effectiveness as well as offering wider social, economic and environmental benefits, that is, de-congestion, reduction in air pollution, healthier lifestyles leading to an overall healthier society and reduced healthcare costs. However, we argue that this emphasis needs to be supplemented with a more critical analysis of the surveillance capacities of these digital technologies. Where in a pre-digital era informational flows were controlled by decisions that were determined by context (Boyd, 2008), the contactless flow of data in times of digital convergence relegates the idea of context to a pure simulation. The example of the Oyster smart card has shown that the data collected in the transport context are readily available for marketing purposes. However, because the user situates both his/her travel activities and associated products such as the Oyster smart card within the framework of a socio-cultural activity (largely thanks to the marketing discourses surrounding ICTs for transport discussed above), the economic and social value of the data collected via these activities is obscured. Thus he/she is not aware of the ways in which this data can then be 'processed, manipulated, traded and used to influence us and to affect our life chances' (Lyon, 2001, p. 108). However, it is precisely our inability to determine exactly how data provided in one context might be used in another, and/or by a different agency and with what consequences, that requires ICTs to be seen as means of surveillance. Thus, the main social, ethical and legal challenges emerging from the use of ICT applications in transport contexts are identified as:

- mission creep;
- profiling;
- privacy;
- corporate exploitation of digital labour.

What follows is a more detailed discussion of these issues, at the end of which we offer a summative, indicative table of the issues that need to be taken into account by stakeholders, giving examples and suggesting potential solutions.

As argued earlier, the collapse of contexts has severe implications for the ability to maintain privacy as it has eliminated the right to draw the boundaries within which one's personal information should remain. This, however, would appear to nullify the principle of 'purpose limitation' which was meant to uphold the very contexts within which the data are collected and for which the user has given his/her consent. As stated in Section I, Article 6 of the EU Data Protection Directive 95/46/EC, 'personal data must be collected for specified, explicit and legitimate purposes . . . and not be further processed in a way incompatible with those purposes'. Returning to the example of the Oyster card, the fact that a user's personal data can be made available to both commercial parties and law enforcement agencies disrespects the principle of purpose limitation. This is what Lodge also refers to as 'mission creep'. As she argues, 'mission creep is unethical because it can be insidiously introduced: if a technological application functions in one context, extending its application to another one – as a "tool" – is tempting, especially if this can be achieved without further reference to public [or] parliamentary scrutiny' (2010, p. 12). In this context it is also important to refer to the debate around what has come to be known within the European Union as 'the right to be forgotten', because the effects of mission creep are exacerbated by the fact that data is often held for unspecified lengths of time. No final resolution had been made within the European Union about this issue up until 2013. However, Viktor Mayer-Schoenberger, one of its strongest proponents, has argued for the introduction of a sell-by date for data, which would mean the deletion of data once it has been used for its primary purpose (*The Guardian*, 2013).

Helen Nissenbaum's argument in the previous section that privacy is contextual becomes thus problematic in so far as mission creep eliminates the boundaries that previously defined what should remain private. Debates around privacy issues are longstanding both in academia and in policy circles with regard to ICTs and surveillance (Bennett, 2012). Corporations in turn have responded more quickly by adapting both their products and communication strategies to this increased sensitivity towards privacy. The CEO of the company 'Renew', for instance, that in August 2013 faced severe criticism for placing data-gathering recycling bins around the Cheapside area of London that captured the MAC codes (Migration Authorisation Code) from the smartphones of passers-by, made it clear his company was taking advantage of a loophole in the law by not adding name or home address to the collected data (Quartz, 2013). However, the name and home address of a person are no longer necessary in order to compile a digital footprint that is information-rich enough to be commercially exploited. Consequently, the notion of privacy itself needs to be redefined. Privacy, as a consequence, is no longer tied to 'individual' variables such as name or home address, but needs to include other variables that are necessarily shared with other members of the population. In this context, to paraphrase Lyon, data is used in ways in which they can 'influence and affect . . . [the] life chances' of entire groups of the population. Hence, the privacy debate needs to move on from the level of the individual to that of the social in order to explore the implications of the widespread use of ICTs – for transport in this case – for social groups.

While the profiling of groups can be directed towards positive ends such as the identification of disadvantages based on disability, for instance, in the context of neo-liberal market capitalism, social profiling becomes a tool of market segmentation. The data collected from ICTs used in transport contexts becomes part of what Pridmore refers to as a 'personal information economy', an economy that 'depends upon the gathering of data through surveillance systems and then analyses this data for patterns and associations deemed to be "of value", continually re-evaluating corporate practices and products based on these analyses' (Pridmore, 2012, p. 321). Where individuals embed ICTs for transport, such as the Oyster card or the PleaseCycle portal and app, within the socio-cultural practice of travel, at the same time they produce monetary value through these practices (Potoglou et al., 2013), thus becoming digital labourers. As Zurawski and Andrejevic have argued, 'it should be noted that watching, monitoring, or profiling actually involve work, constituted of acts and place where surveillance is performed or practiced' (Zurawski, 2011, p. 514). Critical scholars of the media and ICTs are increasingly focusing on the exploitation of digital labour through companies such as Facebook, who utilize user data in order to sell advertising space (Fuchs, 2011). The above considerations show that this is an issue that should become increasingly relevant for transport studies also, as digital technologies are used more and more in this context. For example when applications to calculate the carbon footprints are designed, potential risks need to be taken into account. What the preceding argument highlighted, is that aside from issues around the privacy of persons who submit their personal data in order to benefit from the advantages coming with the increased use of ICTs in transport systems, a range of other ethical, social and legal implications arise that currently receive only limited attention in transport studies. Within the scope of this chapter it was not possible to undertake a full analysis of these implications, yet we suggest that their consideration in the transport studies context would amount to a healthy expansion and redefinition of the field. As with ICTs more generally, debate needs to be fostered with a number of other disciplines, such as law. Depending on whether a given smartphone application makes use of location data services, widely used platforms such as OpenStreetMap or GoogleMaps for instance, may significantly affect data-protection policies that need to be adhered to. Equally, significant differences between countries may be experienced. For example, despite the existence of a common EU policy framework, the national legal frameworks and the implementation of this common EU Directive vary in technologically advanced countries such as the UK and the Netherlands (see Chapter 11). The complexity of the legal and policy framework results in part from the fact that data are collected locally but flow globally. These challenges call for an urgent debate across disciplinary boundaries to enable law and policy responses to new technological developments to more effectively limit the risks involved.

Table 9.1 gives an indicative list of ethical, social and legal issues arising from the increased application of ICTs in the transport context, and offers suggestions about how these issues could be addressed by technology manufacturers and policy-makers. It is not an exhaustive list of issues or possible solutions, but should be understood as an indication of the types of action necessary for a socially acceptable future of transport technology.

Table 9.1 Illustration of some key ethical, social, and legal issues to be taken into account by stakeholders

Issue	Example	Desirable action	Timeline
Purpose limitation/mission creep	Chromaroma: a smartphone app offering discounts for activities conducted by users in their real life e.g. travel by bus	Increase of user awareness about the ways in which one's data, e.g. collected through gamification (PleaseCycle), can be used for other than the originally specified purpose (e.g. when data is made available to third parties for law enforcement or commercial entities)	Issue should be considered during the ICT design process and options offered to users prior to registration
Definition of 'personal information'	The data-gathering bins around London were argued to be 'legal' as the data collected was not deemed 'personal information' as long as it excluded names and addresses	Review of what is considered as 'personal information' taking into account implications of datasets such as time of day not just individually, but also the sensitivities arising for users from potential combination or correlation of individual datasets (e.g. location, time, frequency) which is intertwined with Big Data	Issue should be addressed through appropriate legal frameworks and codes of practice both in advance and during registration to use ICTs
Wider definitional issues	Perception and definition of privacy in ICT initiatives with a community objective e.g. PleaseCycle	Privacy, if understood from an ICT or legal perspective, is largely equated with data protection. In philosophy, however, the history and implications of this concept are far more complex, where it is part of what it means to be human. Decision- and policy-making processes would therefore benefit from more frequent round-table discussions where these conceptual issues can be addressed, as they have significant impact on technology development and implementation	Such issues are of wider relevance and should be reviewed both by policy makers and by users to aid manufacturers/developers to address them efficiently
Quality assurance	Ensure that any key ethical/social/legal issues have been addressed before an ICT is available for public use	Extensive and regular 'ethical reviews' to be conducted in the design and development stages. Stronger cooperation with academic experts or focus groups with potential users could act as independent review processes. Development of a quality standard (e.g. ISO, EMAS) to be adhered is a future alternative	Issue should be addressed at the final stage of ICT development, but should be reviewed throughout the design and development process

6 CONCLUSIONS: THE PLACE OF CRITIQUE IN POLICY-MAKING AROUND ICTs

The practical suggestion here is that there is an urgent need to raise normative questions in the lead-up to policy decisions on ICT applications in a wider sense (see Table 9.1). As already argued, there is a tendency in marketing and policy communications to present ICTs as a neutral means to desirable experiences (Figure 9.1), which follows from the neoliberal market imperative to derive value from the data provided in these contexts. The basic operating principle of digital ICTs has two significant implications: first, that ICTs for transport need to be considered as part of a larger informational infrastructure, rather than as an isolated case, and second, the collapse of the contextual boundaries originally separating the purposes of data collection. This has enabled both commercial entities and law enforcement agencies to benefit from the masses of data gathered through transport applications, often without the explicit consent of the user. Subsequently, a number of ethical, social and legal concerns have been raised, emanating from the increasing violation of the principle of purpose limitation. In this final section, it is argued that critical theories of ICTs and philosophy of technology approaches can inform an interdisciplinary debate on the benefits and risks of ICTs with a twofold purpose: first, to increase the awareness of the public about the risks associated with the use of technologies that collect, process, store and transfer their data in order to be able to make informed decisions. Second, stakeholders from legal and policy-making circles will benefit from a more thorough and theoretically informed debate to avoid a lack of clarity in definitions of concepts such as privacy creating loopholes that can then be exploited for commercial gain, as seen in the above example of the data-collecting bins.

Drawing on the work of the German philosopher Martin Heidegger, who criticizes what he sees as the dominance of an instrumental understanding of technology that conceptualizes technology as a neutral means to beneficial ends (Heidegger, 1977), an understanding can be reached of why users fail to see ICTs as an interconnected informational system and the risks associated with this interoperability. According to Heidegger, the mode of engaging with everyday objects like technologies as 'ready to hand' leads to their usability obscuring their systemic character. From Heidegger's perspective, an ontological analysis of an object like the Oyster smartcard would deconstruct it into a complex network of materials, persons and intentionalities or power relationships. Underneath the apparently simple plastic surface of the card such an analysis would reveal how it is tied into the political economy of personal information, and the network of private, commercial and state actors involved as beneficiaries in its usage. As our prior analysis has shown, use of the Oyster card adds a number of benefits to the socio-cultural practice of travel (Figure 9.1); however, the data collected via this practice is of value to commercial entities and law enforcement agencies, a factor that is part of the very design of the object. A philosophy of technology approach can thus replace a focus on pure usability on the part of the user with a more comprehensive view of the technological object that takes into account the risks arising from its usage also.

At the same time, a comprehensive understanding of technology such as that afforded by philosophy and sociology of technology approaches is needed in the policy-making process (Table 9.1). Currently, a widespread lack of understanding of the technologies themselves, a lack of clarity in the definition of concepts used to regulate their use (privacy being a case in point) and the inadequate consideration of potential risks has led to a situation where the dominant policy strategy is to pass 'corrective legislation'. An example would be the supplementary regulation on bodyscanners at European airports introduced by the European Commission (EC, 2008) following a series of debates on their impact on health and risks to privacy several years after their original trial at London Heathrow, Amsterdam Schiphol and Helsinki Vantaa airports. Contemporary regulations in the US offer the option to travellers not to be scanned, although they are advised that this may result in their missing their flight departure. Both technology- and policy-driven strategies to safeguard human rights and public health regarding the application of ICTs therefore only contribute to risk minimization. As argued in this chapter, it would be possible to anticipate a large part of the social, ethical and legal implications arising from unchecked informational flows (see, for example, section 5) by allowing a philosophically informed debate to take place at an earlier stage in the decision-making process. Thus, it could be ensured that technologies will continue developing 'unchecked by anything other than such policies and regulations that minimize their risks and their misuse' (Kompridis, 2009, p. 21).

To conclude, it should be stressed that, as in other consumer contexts, ICTs for transport do not form part of a 'malign plan of a totalitarian corporation or state' (Zurawski, 2011, p. 515) but have evolved in response to the challenges societies face due to growing urban environments and changing lifestyle patterns. It is in this sense, however, that they need to be considered not in isolation from, but from the perspective of the wider information-technological infrastructure of society. Consequently, the ethical, social and legal implications of the application of ICTs to transport contexts are not just relevant to the transport context but form part of the wider challenges that technological innovation is posing for society.

Thus, while the debates surrounding the controversial data-gathering processes of the Google Street Map View have raised much-needed public awareness in relation to privacy globally, it has been argued that the negative 'hype' created around individual technologies should not distract from the *systemic* nature of the informational infrastructure. It is the purpose of this chapter to highlight some of the ways in which transport studies should expand their field of vision to include some of the more recent debates taking place in critical studies of ICTs such as the exploitation of digital labour. At the same time, these debates should also be enriched by the insights gained from transport research as outlined in the previous chapters.

ACKNOWLEDGEMENTS

The authors wish to thank the participants of the NECTAR Cluster 1 international workshop held in Harrogate in September 2012 for their constructive feedback on an earlier draft of this chapter. Furthermore, we would like to thank Dr Mark Andrejevic at the University of Queensland, Australia, for his detailed feedback on the early stages of the argument developed in this chapter.

REFERENCES

Acquisti, A. and Grossklags, J. (2005) Privacy and rationality in individual decision making, *IEEE Security & Privacy Magazine*, 3(1), 26–33.

Althusser, Louis (1970) Ideology and ideological state apparatuses. Accessed 2 October 2013 at: http://www.marxists.org/reference/archive/althusser/1970/ideology.htm.

Andrejevic, M. (2011) Surveillance and alienation in the online economy, *Surveillance & Society*, 8(3), 278–87.

Andrejevic, M. (2013) *Infoglut: How Too Much Information is Changing the Way We Think and Know*, New York: Routledge.

Bagchi, M. and White, P. (2005) The potential of public transport smart card data, *Transport Policy*, 12(5), 464–74.

Bannister, D. (2012) *Unsustainable Transport: City Transport in the New Century*, Abingdon: Routledge.

Barnes, S. (2006) A privacy paradox: Social networking in the United States, *First Monday*, 11(9).

Bekhor, S., Cohen, Y. and Solomon, C. (2013) Evaluating long-distance travel patterns in Israel by tracking cellular phone positions, *Journal of Advanced Transportation*, 47(4), 435–46.

Bennett, C. (2012) In defence of privacy: The concept and the regime, *Surveillance & Society*, 8(4), 485–96.

Boyd, D. (2008) Facebook's privacy trainwreck: exposure, invasion, and social convergence, *The International Journal of Research into New Media Technologies*, 14(1), 13–20.
Castells, M. (2000) The contours of the network society, *Foresight*, 2(2), 151–7.
Castells, M. et al. (2007) *Mobile Communication and Society*, Cambridge, MA: MIT Press.
Chromaroma (2013) Chromaroma – adding colour to your journeys: A new game for London. Accessed 4 October 2013 at: http://www.chromaroma.com/.
Cottrill, C., Pereira, F., Zhao, F., Dias, I., Lim, H., Ben-Akiva, M. and Zegras, C. (2013) Future mobility survey: Experience in developing a smartphone-based travel survey in Singapore, Proceedings of the Transportation Research Record (TRB) 92nd annual meeting, pp. 1–17.
Dekkers, J. and Rietveld, P. (2009) The adoption of Geo-ICT in economics: Increasing opportunities for spatial research in economics. In Scholten, H., Velde, R. and Manen, N. (eds) *Geospatial Technology and the Role of Location in Science*, GeoJournal Library, 96, pp. 133–61.
Dijst, M. (2009) ICT and social networks: Towards a situational perspective on the interaction between corporeal and connected presence. In Kitamura, R., Yoshii, T. and Yamamoto, T. (eds) *The Expanding Sphere of Travel Behaviour Research – Selected Papers from the 11th International Conference on Travel Behaviour Research*, IATBR, Emerald.
EC (2008) MEMO/08/684: Aviation security: Workshop on Bodyscanners, 7 November 2008. Accessed 4 October 2013 at: http://europa.eu/rapid/pressReleasesAction.do?reference=MEMO/08/684&type=HTML.
EC (2013) Galileo, Europe's GPS, opens up business opportunities and makes life easier for citizens, European Commission, MEMO/13/718, 24 July.
Eriksson, L. and Bjornskau, T. (2012) Acceptability of traffic safety measures with personal privacy implications, *Transportation Research Part F: Psychology and Behaviour*, 15(3), 333–47.
European Commission (2011) Commission Regulation (EU) No. 1141/2011 of 10 November 2011. Accessed 4 October 2013 at: http://eur-lex.europa.eu/LexUriServ/LexUriServ.do?uri=OJ:L:2011:293:0022:0023:EN:PDF.
Fairclough, N. (2001) *Language and Power*, Harlow: Pearson.
Feather, J. (2008) *The Information Society: A Study of Continuity and Change*, 5th edn, London: Facet.
Floridi, L. (2013) *The Ethics of Information*, Oxford: Oxford University Press.
FTC (2012) Mobile apps for kids: Current privacy disclosures are disappointing, Federal Trade Commission staff report, February.
Fuchs, C. (2011) *Foundations of Critical Media and Information Studies*, Abingdon: Routledge.
Garnham, N. (2005) From cultural to creative industries: An analysis of the implications of the 'creative industries' approach to arts and media policy making in the United Kingdom, *International Journal of Cultural Policy*, 11(1), 15–29.
Givoni, M. (2012) Re-assessing the results of the London congestion charging scheme, *Urban Studies*, 49(5), 1089–105.
Google Inc. (2013) iTunes Preview Google Maps. Accessed 4 October 2013 at https://itunes.apple.com/gb/app/google-maps/id585027354?mt=8.
Guardian, the (2013) Right to erasure protects people's freedom to forget the past, says expert, 4 April 2013. Accessed 4 October 2013 at: http://www.theguardian.com/technology/2013/apr/04/right-erasure-protects-freedom-forget-past.

Guiry, J., van De Ven, P. and Nelson, J. (2012) Orientation independent human mobility monitoring with an Android smartphone, Proceedings of the IASTED International Conference on Assistive Technologies (AT'12), ACTA Press, pp. 800–808.
Haggerty, K.D. and R.V. Ericson (2000) The surveillant assemblage, *British Journal of Sociology*, 51(4), 605–22.
Heidegger, M. (1962) *Being and Time* (trans. J. Macquarrie and E. Robinson), Oxford and Cambridge, MA: Blackwell.
Heidegger, M. (1977) *The Question Concerning Technology and Other Essays* (trans. William Lovitt), New York, London, Toronto, Sydney: Harper.
HMC (2013) Improving local transport. Accessed 4 October 2013 at: https://www.gov.uk/government/policies/improving-local-transport/supporting-pages/smart-ticketing.
Horowitz, J. and McConnell, K. (2002) A review of WTA/WTP studies, *Journal of Environmental Economics and Management*, 44(3), 426–47.
Ideas in Transit (2012) List of innovations. Accessed 4 October 2013 at: http://www.ideasintransit.org/wiki/List_of_innovations.
IWGDPT (2010) Mobile processing of personal data and security, International Working Group on Data Protection in Telecommunications working paper 675.41.18, 48th meeting, 6–7 September, Berlin.
Jenkins, H. (2006) *Convergence Culture: Where Old and New Media Collide*, New York: NYU Press.
Kanngieser, A. (2013) Tracking and tracing: geographies of logistical governance and labouring bodies, *Environment and Planning D: Society and Space*, 31(4), 594–610.
Kittler, F. (1999) *Gramophone, Film, Typewriter* (trans. G. Winthrop-Young and M. Wutz), Stanford: Stanford University Press.
Kompridis, N. (2009) Technology's challenge to democracy, *Parrhesia*, 8, 20–33.
Kosinski, M., Stillwell, D. and Graepel, T. (2013) Private traits and attributes are predictable from digital records of human behaviour, *Proceedings of the National Academy of Science of the United States of America*, 110(15), 5802–805.
Lodge, J. (2010) Quantum surveillance and 'shared secrets'. A biometric step too far? *CEPS Liberty and Security in Europe*, Brussels.
Lyon, D. (2001) *Surveillance Society: Monitoring Everyday Life*, Buckingham: Open University Press.
Lyon, D. (2007) *Surveillance Studies: An Overview*, Cambridge: Polity.
Lyons, G. and Urry, J. (2005) Travel time use in the information age, *Transportation Research Part A: Policy and Practice*, 39(2–3), 257–76.
Lyotard, F. (1984) *The Postmodern Condition*, Manchester: Manchester University Press.
Meyrowitz, J. (2008) Medium theory. *Blackwell Reference Online*. Accessed 15 January 2014 at: http://www.blackwellreference.com/public/tocnode?id=g9781405131995_yr2013_chunk_g978140513199518_ss65-1.
Midgley, P. (2009) The role of smart bike-sharing systems in urban mobility, *Journeys*, 2, 23–31.
Mokhtarian, P. and Salomon, I. (2001) How derived is the demand for travel? Some conceptual and measurement considerations, *Transportation Research Part A: Policy and Practice*, 35(8), 695–719.
Mott MacDonald (2009) Productive use of rail travel time and the valuation of travel time savings for rail business travellers, London: Department for

Transport. Accessed 1 May 2013 at: https://www.gov.uk/government/uploads/system/uploads/attachment_data/file/4003/productive-use-of-travel-time.pdf.
Murakami Wood, D. and Ball, K. (2013) 'Brandscapes of control: surveillance, marketing and the co-construction of subjectivity and space in neo-liberal capitalism', *Marketing Theory*, 13(1), 47–67.
Nakamura, K. and Hayashi, Y. (2012) Strategies and instruments for low-carbon urban transport: An international review on trends and effects, *Transport Policy*, 29, 264–74.
Nissenbaum, H. (2004) Privacy as contextual integrity, *Washington Law Review*, 79(1), 119–58.
Owen, R., Baxter, D., Maynard, T. and Depledge, M. (2009) Beyond regulation: Risk pricing and responsible innovation, *Environmental Science & Technology*, 43, 6902–906.
Potoglou, D., Patil, S., Gijón, C., Palacios, J. and Feijóo, C. (2013) The value of personal information online: Results from three stated preference discrete choice experiments in the UK, Proceedings of the 21st European Conference on Information Systems, 5–8 June 2013, Utrecht, the Netherlands.
Pridmore, J. (2012) Consumer surveillance: Context, perspectives and concerns in the personal information economy, in Ball, K., Lyon, D. and Haggerty, K. (eds) *Routledge Handbook of Surveillance Studies*, London: Routledge.
Quartz (2013) This recycling bin is following you. 8 August. Accessed 4 October 2013 at: http://qz.com/112873/this-recycling-bin-is-following-you/.
Rose, G. (2006) Mobile phones as traffic probes: Practices, prospects and issues, *Transport Reviews*, 26(3), 275–91.
Schaffer, H., Komninos, N. and Pallot, M. (2012) Smart cities as innovation ecosystems sustained by the future Internet, Fireball white paper. Accessed 8 September 2013 at: http://hal.inria.fr/docs/00/76/96/35/PDF/FIREBALL-White-Paper-Final2.pdf.
Steenbruggen, J., Borzzachiello, M-T., Nijkamp, P. and Scholten, H. (2013) Mobile phone data from GSM networks for traffic parameter and urban spatial pattern assessment: A review of applications and opportunities, *GeoJournal*, 78(2), 223–43.
TfL (2012) London buses factsheet. Accessed 12 December 2013 at: http://www.tfl.gov.uk/assets/downloads/corporate/london-buses-factsheet.pdf.
TfL (2013) Facts and figures. Accessed 4 October 2013 at: http://www.tfl.gov.uk/corporate/modesoftransport/londonunderground/1608.aspx.
Thakuriah, P. and Geers, G. (2013) Technology systems for transportation system management and personal use, in *Transportation and Information: Trends in Technology and Policy*, Heidelberg: Springer, pp. 35–71.
Transport for London, (n.d., a) What is Oyster? Accessed 4 October 2013 at: http://www.tfl.gov.uk/tickets/14836.aspx.
Transport for London (n.d., b) Oyster card. Accessed 4 October 2013 at: http://www.tfl.gov.uk/termsandconditions/12321.aspx.
Tyrinopoulos, Y. (2011) Integrating vehicle positioning data in quality control programs in public transit: A European perspective, *Transportation Planning and Technology*, 34(2), 167–76.
UNCRC (1989) Convention on the Rights of the Child – Adopted and opened for signature, ratification and accession by General Assembly Resolution 44/25 of 20 November 1989, entry into force 2 September 1990, in accordance with Article 49.

Van Wee, B. et al. (2006) Is average daily travel time expenditure constant? In search of explanations for an increase in average travel time, *Journal of Transport Geography*, 14(2), 109–22.

Waze (2013) Get the best route, every day, with real-time help from other drivers. Accessed 16 January 2014 at: https://www.waze.com/.

Webster, F. (2002) *Theories of the Information Society*, 2nd edn, London: Routledge.

Xiao, Y. et al. (2012) Transportation activity analysis using smartphones, Proceedings of the IEEE Consumer Communications and Networking Conference (CCNC), Las Vegas, 14–17 January 2012, pp. 60–61.

Xynou, M. (2012) Oyster, Octopus and Metro cards: What happens to our data? Privacy International, 27 October. Accessed 4 October 2013 at: https://www.privacyinternational.org/blog/oyster-octopus-and-metro-cards-what-happens-to-our-data.

Zurawski, N. (2011) Local practice and global data: Loyalty cards, social practices, and consumer surveillance, *The Sociological Quarterly*, 52, 509–27.

10. Intelligent Transport Systems: glocal Communities of Interest for technology commercialization and innovation

Margarita Angelidou, Nicos Komninos, Xavier Leal, Isidoros A. Passas, Maria S. Schoina and Elena Sefertzi

1 INTRODUCTION

Background and Problem Description

Information and Communication Technologies (ICTs) have had a tremendous impact on industrial and transport sectors. The problem of traffic management in densely populated urban agglomerations has become a main policy focus of forward-looking local administrations seeking to employ innovative practices, offer improved services to their citizens and, ultimately, foster competitiveness and attractiveness. In addition, since 2005 the rise of the Web 2.0 has transformed users from passive receivers of information to active content generators, thus enabling their organization in virtual communities. Users are now having their say, and the successful development of ICT services and products depends largely on public acceptance and uptake. In this context the degree and speed of commercialization of information products and services, including Intelligent Transport Systems (ITS), is largely affected by different actors and their communities as well as the way they use these products within their complex networks of interaction.

More specifically Communities of Interest (CoI) represent loose networks of people assembled around a topic of common interest. Their members take part in the community to exchange information, obtain answers to questions or problems, improve their understanding of a subject, share common passions or just play. Their synergy, however, is of a lower level than that of a formal group which is motivated by a common goal. In a glocal CoI, virtual networking can play a role in innovation similar to physical communities in terms of communication, socialization and learning. As is the case with physical communities, the social context of the emergence of the community as well as the strengthening of bonds and intentionality constitute the reference points for describing the characteristics of glocal communities. This chapter addresses the operation of CoI in the ITS industry. Following the introduction, sections 2, 3 and 4 analyse the landscape of innovative ITS products and services, and selected case studies and commercialization routes of ITS products and services in Greece are presented. Finally the conclusion discusses the way to accelerate commercialization of state-of-the-art ITS products and services through CoI, including a roadmap for the successful launch of new ITS products and services to potential markets.

Taking into account the aforementioned, this chapter was written with the purpose of examining how *CoI, particularly glocal, can reinforce innovation in the ITS sector*. In effect this chapter has three objectives:

1. to achieve a thorough review of state-of-the-art ITS technologies;
2. to analyse how the commercialization of innovative ITS products and services took place in selected Greek cases, by describing the actors and the routes that were followed to this end;
3. to draw conclusions about how glocal CoI can enhance innovation in the ITS sector and provide a respective roadmap for the commercialization of innovative ICT products.

This chapter is based on findings of the EC FP7 research project entitled 'Enhancing the transfer of Intelligent Transportation System innovations to the market' (T-TRANS), coordinated by the Universitat Autònoma de Barcelona (http://www.T-TRANSnetwork.eu/T-TRANS/). The aim of T-TRANS is to explore the mechanisms that facilitate the transfer of innovative ITS products and services to the market as well as the routes that are used to this end.

Communities of Interest

According to Armstrong and Hagel (2000), CoI bring together participants who interact extensively about a specific topic of interest; these participants interact among each other within the very frame of this specific topic area. In essence CoI bring together individuals from different cultures and disciplines, and their major unifying feature is the common concern to resolve the same problem. ITS members take part in the community to exchange information, obtain answers to personal questions or problems, improve their understanding of a subject, share common passions or play (Henri and Pudelko, 2003). A CoI actually brings together stakeholders from different Communities of Practice (CoPs), that is, groups of practitioners from different domains, and as such it can be thought of as a 'community of representatives of communities' (Fischer, 2001a). However, this diversity of CoI members is both a challenge and an asset in the framework of its function: a challenge because the fact that they come from different domains and diverse cultures forges a communication divide among them, and thus reaching a common understanding can be very challenging; an asset because their interdisciplinary mix yields different insights and uplifts the result of collaboration to a whole new level of effectiveness (Fischer, 2001b).

CoI typically encompass a wide range of stakeholders with members deriving from authorities, universities, development centres, business support organizations, science parks, associations of entrepreneurs, non-profit and cultural organizations, and other relevant actors. They can be citizen representatives, merchants, researchers, software developers, end users and so on. Their segmentation into categories on the basis of their common features is clearly a matter of the viewpoint that is adopted. For example, CoI can be categorized according to their local or glocal outlook, the inclusion or exclusion of experts and so on. Fischer (2001a) gives some examples of CoI: (1) a team interested in software development that includes software designers, marketing specialists, psychologists and programmers; or (2) a group of citizens and experts interested in urban planning who are concerned with implementing new transport systems.

Glocal CoI are of particular concern in this chapter. The glocal nature of CoI denotes that they are global and local at the same time, that is they combine the needs and impacts deriving from globalization with those of local considerations. Although a local dimension has always been present in CoI, their global dimension was enabled and facilitated under the influence of modern ICTs sometime after 2005. This dimension coincided with the diffusion of Web 2.0 technologies, thus enabling interactions beyond the physical boundaries of traditional communities. Web 2.0 technologies have thereupon allowed the formation of virtual communities and the creation of user-generated content. Social media such as Twitter, Facebook, Google+ and wikis have made this transition possible, as they enhance three primary functions that we already have in our communities (Bacon, 2012):

1. *broadcasting* of events, developments, content and other things going on in a community;
2. *feedback* from community members, users, customers and other people;

3. *collaboration*, collective event and project planning, collaborative campaigns and efforts.

Not only did these changes cultivate the need to organize in communities, they also provided the tools to these communities both to expand their reach, by enabling their members to communicate and collaborate over long distances, and to take new forms based on shared practice (Wenger, 1998).

The operation of CoI could not have left the commercialization processes of innovative ITS products and services unaffected. The rise of the information society, the global spread of the Internet, and the creation of virtual networks along with community digital spaces have opened new ways for the functioning of innovation supply chains. Formerly free-standing information products are now combined and offered in platforms (Wenger et al., 2005). At the same time new innovation tools emerged, enabling strategic intelligence and market watching, online technology learning, online collaboration on new product development, online marketplaces and the promotion of products to global markets. All these developments have reinforced cooperation possibilities among actors operating in the transport sector.

2 INTELLIGENT TRANSPORT SYSTEMS: STATE-OF-THE-ART TECHNOLOGIES

The transport of today and the future relies heavily on ITS for numerous reasons related to: (1) environmental protection, by monitoring carbon emissions and carbon footprint; (2) delivering fast and reliable services, thus providing users with safe and comfortable journeys; and (3) transport economics regarding cost, new products, markets and businesses. The large number of ITS applications and technologies, either already implemented, just introduced or even in the early development stages, indicates their importance to the transport industry while the industry itself is one of the main 'ITS customers'.

The ITS concept currently incorporates numerous technologies and applications, under various thematic areas, and spanning different transport modes although they mostly originate in the road transport mode for individualized transport. The term can be found in the literature with numerous definitions, the more detailed given by the European Telecommunications Standards Institute (ETSI, 2012): 'Intelligent Transport Systems use information and communication technologies to improve safety, reliability, efficiency and quality of transport. The technologies are therefore added to infrastructures and are integrated at the corresponding transport vehicles.'

An analysis of the ITS domain map is presented below based on the four most representative ITS domains, under the criterion of presenting a strong market vision interest. These domains are:

1. smart grid: connection, charging and storage of energy;
2. revenue management technologies for freight transport;
3. intelligent Intermodal Transport Units;
4. rail networks: a technological system-wide approach.

These areas include plenty of elements structured in different levels, which were identified in this review as technologies and applications. Their position and use is not often distinct since, on the one hand, an application may include multiple technologies, while on the other hand a certain technology can be found in different applications, and all these can be integrated to more than one domain.

Smart Grid

The smart grid, from the ITS point of view, focuses mainly on Electric Vehicles (EV). The automotive industry is rising globally while environmental restrictions and measures, implemented by many countries worldwide and the European Union (EU), drive the EV penetration in the market. However, according to Accenture (2011), for a large-scale roll-out of EVs, investors should stimulate demand through changing consumer behaviour, lowering the cost of public charging infrastructure and managing the grid impact. ICTs can play an important role in this direction through the creation of intelligent charging systems, by providing real-time data exchange between the vehicle and the grid as well as by reducing residual load fluctuation.

Paradigms in the US, such as eVgo and SFpark, demonstrate that open innovation can overcome the barriers to broader adoption of EVs in the market. Finally the main problem with EVs is the standardization of different components: plugs, charge stations, paying, cables and so on. In Europe several initiatives have sprung up attempting to standardize different parts of the EV infrastructure. The relative applications, based on one or more technologies, fall into the categories below:

1. availability check;
2. EV booking and payment;
3. EV charging;

4. fleet management system;
5. grid management;
6. Vehicle-to-Grid (V2G);
7. Vehicle-to-Infrastructure (V2I).

Revenue Management

For several years now air transport has been closely connected to the ICT sector, far more than any other transport mode. Complex ICT systems, such as revenue management techniques, have been used in passenger airlines for decades, but in the field of commercial air transport they have been in use only over the last 15 years. The main reason for this lies in differences such as capacity, demand, space allocation and routing existing between the two businesses, which, mathematically speaking, makes the cargo revenue management problem far more difficult to solve than the passenger one (Talluri and van Ryzin, 2005).

Air cargo is a growing market accounting for 12 per cent of all industry revenues (IATA Annual Review, 2013). The potential for its efficient exploitation relies heavily on breaking through revenue management systems that need to address complex calculations for the optimization of transport. Furthermore recent ITS solutions are trying to address important issues of the air cargo industry, such as shipment documentation leading to intelligent cargo and e-freight initiatives, communication between supply chain actors using software-as-a-service/cloud computing models and airports' ground operations adopting radio-frequency identification.

A lot of functions take place in this domain, and ICTs play a critical part in intelligent applications. The main applications identified here are:

1. capacity forecasting;
2. demand forecasting;
3. overbooking;
4. capacity management;
5. freight transport booking;
6. freight tracking;
7. airport ground operations.

Intelligent Intermodal Transport Units

Intermodal freight transport has a close relationship with the ICT sector, and has a lot more to gain from it in the future as it can provide the framework for addressing new challenges. Intelligent Intermodal Transport

Units (ITUs) are considered to be highly promising for an efficient, productive, profitable, safe and secure interrelated distribution process, addressing the requirements for low environmental impact and strict quality control. Furthermore recent technological developments (agents, service oriented architectures, Internet of Things) have provided the means for the development of related applications in the logistics sector. Conversely, current applications on the internal monitoring system of ITUs are mostly based on closed platforms and aim at niche requirements rather than the sector as a whole.

Nevertheless there is a significant gap between research and innovation, on the one hand, and the actual application, on the other, since none of the developed solutions is being implemented in large-scale intermodal transport systems globally. The doubtful Return on Investment (ROI) in state-of-the-art technologies, the unwillingness to share data, the emergence of day-to-day logistic bottlenecks, difficulties in matching business and environmental protection, non-interoperable devices and platforms and insufficiently decentralised decision-making are some of the reasons for this situation. Collaboration among shippers, carriers and service providers, and optimized solutions and innovations harmonize standards involving main standardization bodies, thus decreasing the barriers in the Intelligent Intermodal Transport Innovation Chain. Either in pilot or large-scale applications, the following applications rely greatly on ITS:

1. electronic registration/single window systems;
2. access control;
3. e-seals;
4. automatic vehicle location;
5. ITU2ITU/V2ITU;
6. freight tracking;
7. condition monitoring;
8. remote freight information.

Rail Networks

The goal of the railway industry is to maximize the capacity of existing networks as traffic, both passenger and freight, has been growing rapidly and the construction of new tracks is most likely unfeasible. Capacity and punctuality, therefore, need to be addressed through advanced information technology and control systems. The latter reduce the impact of human factors and other unpredictable parameters by identifying them in real time and updating or modifying operations. Thus rescheduling and controlling are the two functions that need to be implemented successfully, and the requirements are high precision of the train position, data of the train features and accurate traffic prediction, based on predefined criteria. Early applications show that this approach can significantly improve service quality even as additional trains are added to the timetable.

Train Control Systems (TCSs) differ from country to country. Currently there are 20 TCSs across the EU. To solve this problem, the 'European Rail Traffic Management System' initiative foresees the replacement of all systems in Europe with a single advanced, compatible and interoperable European TCS, with significant benefits for the railway sector. Non-EU countries though, such as China, India and Taiwan, are adopting it as well, acknowledging its advantages.

While the respective requirements are not entirely met by complete systems in the market, parts of their subsystems already feature most of these requirements. Some main applications that can be distinguished are:

1. driver assistance systems;
2. collision avoidance;
3. automatic vehicle location;
4. computer aided dispatching;
5. safety information exchange;
6. traffic management system;
7. vehicle diagnostics.

The CoI concept as described in the introduction of this chapter is connected to the aforementioned in different ways. In some cases the contribution of stakeholders or participants can be of major impact in the deployment of ITS solutions, though in other cases this interaction and a bottom-up approach might not be efficient enough for an innovation process.

The electric vehicles ecosystem, for example, can be highly related to CoI as drivers, electricity providers, grid managers, fuel and charging station managers, and equipment manufacturers could interact, having a positive impact on issues related to smart grid and EVs. On the other hand revenue management solutions for air cargo transport mainly depend on developers, mathematicians and ICT experts in order to address the complicated situation in this domain.

Rail transport operations can benefit a lot from the integration of passengers in the real-time information system, along with the data provided by drivers, guards, transponders and other trackside equipment. Thus rescheduling and controlling processes can be executed more efficiently for the punctual and safe operation of the rail network. As for the intermodal transport units, the establishment of CoI of the numerous stakeholders involved (even in one single transport) can be of great impact in overcoming the barriers occurring for the integration of ICT in this domain. The devices necessary for these applications could get smaller and cheaper, standardization and interoperability could be obtained and unwillingness of operators to use them could be diminished.

3 A SURVEY ON ITS CASE STUDIES IDENTIFYING COMMERCIALIZATION ROUTES

The commercialization of innovative products and services is a constant pursuit, both in academia and the enterprise sector, as it is one of the ways to help companies gain competitive advantage. In the framework of the T-TRANS project, a mapping exercise was performed in the greater area of Thessaloniki to identify existing ITS initiatives. The selection of these initiatives/applications was made in accordance with the main scope of the T-TRANS project, which is to enhance the transfer of innovative ITS solutions and applications to the market. To this end eight initiatives related to ITS were analysed, and 21 potential CoI members were identified. The identified applications/initiatives are (T-TRANS, 2013a):

1. smart commercial freight gates on ports;
2. Thessaloniki's intelligent Urban Mobility Management System;
3. the OASTH Telematics Passenger Information Project;
4. Municipality of Kalamaria Mobility Centre in Thessaloniki;
5. Taxibeat;
6. parking defenders;
7. freight rail tracking;
8. SMART-TAXI.

Three of the initiatives are focused on public transport, one on private transport in metropolitan areas, one on freight transport in commercial sea ports, one in parking and one on rail freight transport. All these initiatives are currently in operation serving millions of users every day. Four of the initiatives have been implemented by legal entities of public law, not all of them being strictly public entities, while three of them are private initiatives. Most of the initiatives have been co-funded by national resources and the European Commission (EC), either directly or indirectly and in varying co-funding rates.

The first commercialization route that was identified is to build the business case of the product or service utilizing funding opportunities, offered either by national or European funding schemes. National or European co-funding schemes provide great support to both academia and enterprises for Research and Development (R&D), with the purpose of designing innovative products or services to satisfy user needs. As the analysis of the initiatives showed, the drawback which emerges for this route is that most of the initiatives are being implemented as definitive projects, with minor consideration of the further exploitation of the produced product or service.

In the 'traditional approach', five developmental components of a project can be distinguished, four stages plus control to be exact: initiation; planning and design; execution and construction; monitoring and controlling systems; and completion. In this first route, the initiation usually comes from a call for project proposals published by the EC, while the planning and design comply with the guidelines of each call. The monitoring and controlling of the project are mainly based upon the rules set by the EC, while the completion phase has to comply with the proposal and usually takes the form of a set of deliverables to the EC. The commercialization of the product or service in most of the cases finishes with the end of the project life cycle.

Applying typical entrepreneurial market processes is the second identified commercialization route. The market domain of ITS products and services can be divided into the three traditional market categories, namely, business to business (B2B), business to consumer (B2C) and business to government (B2G). The new product development process is initiated by either an idea-generating step or the identification of a new market need. One of the characteristics in this route is that the new products emerge under a short-sighted perspective, within a framework that is set by the accumulated experience as well as by the existing formal and tacit knowledge of the enterprise.

The analysis of the initiatives showed that only one of them was initiated as an entrepreneurial market process. Having identified a business opportunity and using mainly its own financial resources, the company designed and developed an ITS solution for a specific market segment. What was interesting in the examined case was that a couple of years ago the company, after realizing that its product was entering its maturity phase and needed a major upgrade, decided to involve, in the design of the new version of the product, a cluster of enterprises that were also potential customers of the production solution.

The third route is the commercialization through the active involvement of a group of users or clusters, potential stakeholders and beneficiaries. The typology of this route is relatively new, and a great push for its adoption by the enterprises was given with the full deployment of Web 2.0 sites and applications. Web 2.0 describes websites that use technology beyond the static pages of earlier websites. The term was coined in 1999 by D. DiNucci and was popularized by Tim O'Reilly at the O'Reilly Media Web 2.0 conference, in late 2004. Some of the key principles of Web 2.0 are that users add value, network effects take place by default, and the data on a website is the Next 'Intel Inside' (O'Reilly Media, 2005).

In this route small and typologically different pieces are loosely joined. These pieces are the users or the clusters, the potential stakeholders and beneficiaries that will benefit from the new product/service. Utilizing mainly web-based networking tools and thus rendering real space irrelevant, the above 'pieces' can be either loosely or strongly joined and add knowledge and information to the whole commercialization route. The participation of the above actors appears to be diminishing the previously described short-sighted perspective of the internal new product development framework. Of all the initiatives examined, two involve groups or clusters of users and enterprises in the commercialization process. In this third route, the available product/service appears to be state-of-the-art, to achieve greater financial sustainability and to be more democratically developed. Potential customers and other interested stakeholders know from the beginning that its features will satisfy their needs.

The commercialization of a new product or service is a tedious process and the above three routes are not mutually exclusive, since a combination of them can be encountered as well.

As already defined in the first section of this chapter, a CoI actually brings together stakeholders from different Communities of Practice (CoPs), that is, groups of practitioners from different domains, and as such it can be thought of as a 'community of representatives of communities' (Fischer, 2001a). In two of the identified applications/initiatives in Thessaloniki and all over Greece, the participation of different stakeholders has been identified as CoI. In 'Taxibeat' the interests of two stakeholder groups are merged so that both of them satisfy their own needs. The first group of stakeholders consists of the taxi drivers while the second group consists of the taxi passengers. The Taxibeat application merges these two groups into a CoI. Taxi drivers upload their profile and allow the application to track the position of their vehicle and display it on a real-time map continuously. Taxi passengers, by rating each taxi driver, help the CoI to choose the best taxi driver. The CoI, with the support of ICT, that is, the Taxibeat application and GPS tracking, assist both passengers in choosing the best taxi and taxi drivers in expanding their business from a pool of potential customers.

In the second application, 'SMART-TAXI', the CoI has been developed by the merger between a privately owned ICT and software development enterprise and a taxi transport company in Thessaloniki. The latter currently has 2500 members and a fleet of 1200 privately owned vehicles. The ICT enterprise has already been selling an application for the better organization of Taxi companies for quite a few years. Eventually it did realize that the application should be upgraded and new features should be added so that it covers more needs of taxi drivers and taxi owners. The CoI that emerged from the amalgamation of the two different stakeholder groups helped the software company to design and implement new features in the SMART-TAXI application, and, furthermore, the taxi drivers and owners to fulfil their specific needs and requirements that otherwise would have been difficult to identify by the software company alone.

4 ENHANCING INNOVATION IN ITS THROUGH GLOCAL COI

Current Innovation Trends

The Organisation for Economic Co-operation and Development (OECD) Innovation Strategy (2010) offers a good overview of current innovation trends that affect commercialization of products and services. In the first place the Strategy refers to the latest (3rd) edition of the *Oslo Manual*, which distinguishes three types of innovation with respect to different degrees of novelty: (1) innovation-new-to-firm, which corresponds to the diffusion and absorption of existing technologies already implemented by other firms; (2) innovation-new-to-market, also existing solutions introduced by a firm to its market; and (3) innovation-new-to-world, presented for the first time in all markets and industries. Most innovations belong to the first category as the innovation diffusion curve spreads out novelties to the entire population of companies. Then the strategy outlines a series of processes and factors which shape innovation, as it occurs today, beyond the supply-side drivers of R&D and technology:

1. Science and research continues to be an essential driver of innovation; however, many highly innovative firms do not engage in R&D at all, and the majority of firms introducing new-to-market innovation do perform in-house R&D.
2. As the complexity, costs and uncertainty of innovation increase, firms turn to collaboration and partnership to mitigate those risks. Firms turn to open innovation and collaborative networks, especially the formation of global networks, and external knowledge partnerships for the acquisition of knowledge, fresh ideas and market access.

3. Users and consumers also have a growing role, increasing the interaction between demand and supply. User ideas and experience with products can help companies in shaping future innovations and design products that satisfy their needs better.
4. Multinational firms have a leading role in the globalization of innovation. New important players have also appeared in the global landscape. The rapid growth in the People's Republic of China has led to a spectacular increase of R&D expenditure and triadic patents.
5. Local knowledge and capabilities as well as proximity to research and education institutions continue to matter for innovation. Some regions are global leaders in knowledge-intensive activities, and half of the R&D performed in the OECD countries takes place in 10 per cent of the respective regions.

These trends indicate clearly that there is no return to the old linear model of innovation, and R&D translates directly and spontaneously to innovation. On the contrary, the systemic and network perspective is consolidated and shapes all drivers of innovation creation, such as universities and tertiary education, patenting and technology transfer, knowledge infrastructure and flows, international cooperation, governance and stakeholders' involvement in shaping policies for innovation.

The network paradigm of innovation adopted by the OECD is also consolidated in the literature by observing the evolution of innovation drivers over the last 20 years: from R&D commercialization to institutional and territorial systems and networks, to open innovation, to user-driven innovation and to innovation in smart networks and environments. In the network perspective, innovation is the outcome of a nexus of collaboration created between an innovation company and organizations involved in R&D, information dissemination, market making, technology transfer and adaptation, funding, customers and end users (Radjou, 2004; Carayannis et al., 2007; Komninos, 2008; McQuivey, 2013). Such small networks comprising only a few nodes evolve to become extremely large networks with hundreds of participants from all over the world. They include local and global partners, but with the spread of ICTs and virtual networks they are becoming glocal, combining local competences with global know-how and access to markets.

Innovation in the ITS Sector: Lessons from T-TRANS Project

The T-TRANS domains described in the first section bring to the surface some key characteristics of innovations in the ITS sector. Clearly the majority is type 1, 'innovations-new-to-firm', introduced by transport organizations, which have already been implemented and tested in other organizations and regions. End-users of these innovations are transport agencies, public or semi-public authorities and, in all cases, organizations regulated by public institutions. The technological capability and know-how of these ITS end-users are low, they are mere technology absorbers relying on external technology providers for solution development and implementation. We have also seen that ITS solutions rely on a large mix of ICTs, which can be applied to various transport modes. Technologies are combined into dedicated applications, and the ontology of ITS captures the extent of the wide range of applications over various fields and types of services. This complexity is manifested in the collaborative network that introduces innovations, which usually comprises a global leading company developing the technology, a local marketer, a transport consultancy, a local ICT provider and state agencies for funding and regulation.

This type of innovation process and the usually small closed implementation network suffer from a number of weaknesses, which can be primarily attributed to limited strategic intelligence of end users (transport agencies/public organizations) over ITS technologies and innovations. The limited horizon scanning capabilities of transport organizations, for *ex-ante* assessment of alternative ITS technologies and applications, subsequently lead to discontent about the quality of services offered, insufficient estimation of maintenance costs and after-service support as well as failure of the business model, which assures the long-term sustainability and operation of ITS solutions.

However, these weaknesses seem alleviated in cases of more extended innovation supply chains and wider networks. Intelligence is clearly improved as a larger number of actors are involved, and a better understanding of eventual shortfalls is achieved with open consultation and larger user participation. The positive relationship between ITS sustainability, collaboration and partnership is interpreted as an indication of positive impact mainly of online CoI upon ITS innovation, and sustains the arguments for greater involvement of glocal CoI in the ITS sector.

Putting CoI in Action: A 'Triple A' Roadmap

Bacon (2012) in *The Art of Community* presents a detailed framework about the setting and governance of online communities and discusses processes of empowerment and leading communities to success, such as building an effective community strategy meeting stakeholder requirements; creating excitement and participation as underlying motives for achieving community goals; using social media to raise awareness and active involvement; organizing community events; governing infrastructure; selecting community managers and learning from the best established community leaders. Communities are fundamentally organized by non-profit practices, and their participants build up social capital via contributions, interaction and trust. The sense of 'belonging' is critical in activating communities, and 'belonging' is the goal of community building and a measure of the social capital that keeps it alive. Online communities have to be created and sustained on solid ground on the basis of mission and goals, areas of collaboration, skills and needs required. Apart from these initial steps, communities need continuous governance, which is to lead, initiate and engage the members in activities that affect the community as a whole.

Communities, in the field of innovation, should always be glocal to offer state-of-the-art know-how, frontier research outcomes, awareness of local needs and local implementation capabilities. Furthermore glocal communities are online communities because media and web applications are necessary to bring together participants from all over the world.

Working with online user and lead user communities for the design and the development of smart environments and ITS applications, in the framework of the CIP Smart City project 'PEOPLE' (http://smartcityapps.urenio.org/improve-my-city_en.html), researchers from URENIO (AU of Thessaloniki Research Unit) developed a 'triple A' roadmap for the activation of 'Actors', development of 'Architectures' of networking and undertaking innovation 'Activities'. This roadmap can serve as a model and can be further adapted for engaging glocal CoI into ITS solutions.

1. Actors' activation

Identification of actors is the starting point of the roadmap and it is made with respect to the purpose and the objectives of the community. For each type of smart environment or ITS application, it is necessary to identify stakeholders, potential participants and users, and information should be gathered about contact details, interest and motivation to participate, involvement strategy, ethics and confidentiality. Depending on the application, communities can include professionals, R&D organizations, service producers, merchants, public administration personnel, non-profit and cultural organization representatives, researchers, software developers and end users. In principle, three different categories of participants should be chosen: those sustaining the offer of the application/solution, those involved in the demand side, and those in the marketing and dissemination of the service.

Mobilization of communities is the second step and takes place through selected individuals and community leaders. A great deal has been written about the role of social media, such as Facebook, Twitter, LinkedIn, Yammer and others, in creating and mobilizing online communities. However, from our experience, the contribution of social media to creating and mobilizing communities is useful but not as decisive as the role of key individuals who spread the word, encourage community members to get involved and build momentum around the vision of the community. Open source communities and infrastructures such as GitHub, SourceForge and Joomla are also very effective in mobilizing the developer community.

The *contribution* of communities to applications and solutions takes place bottom-up by their members. Under no circumstances are communities involved as organized groups. On the contrary, members, users, producers and citizens contribute in many different ways: giving feedback to applications and e-services; improving the application concept and design, and validating solutions and services; taking part in the implementation of services as producers or consumers; providing content, information, requests and other digital material necessary for the uptake of services; sustaining e-services and online applications in the long run with skills and resources, thus taking part in the implementation of business models for sustainability.

Extensive user involvement increases the probability of successful solutions. A good combination is working with open source, involving large communities of developers and employing widely used development platforms resulting in solutions capable of high diffusion and penetration into world markets. Eventually it is the involvement of users that makes the difference.

2. Architectures of networking – knowledge architectures

The activation of communities sets in motion a distributed knowledge system. The main elements of this system are: transport organizations, technology providers, people and other users; transport infrastructures and physical spaces; embedded devices such as sensors and actuators; software applications; operators and utility organizations. Architectures of networking are knowledge architectures which define how these elements are organized, how information moves from one element to the other and how information is transformed with the collaboration or presence of other elements. Each and every e-service offered by ITS applications relies on a specific knowledge architecture, which brings together a number of constituting elements and determines information flow and knowledge generation.

The main modelling tools commonly used to define data models and knowledge architectures are UML Graph design and RDF/S ontology design. Unified Modelling Language (UML) is a standardized general-purpose modelling language in the field of object-oriented software engineering. UML includes a set of graphic notation techniques to create visual models of object-oriented software-intensive systems. It combines techniques from data modelling (entity relationship diagrams), business modelling (work flows), object modelling and component modelling. Moreover it is proven as an efficient tool to handle complex data models and data warehouses (Prat et al., 2006). A UML diagram is a partial graphic representation of a system's model, while the UML model also contains documentation that drives the model elements and diagrams, such as written cases of use. Additionally the RDF/S ontology design can be used to describe metadata that are associated with the data classes defined in UML. Information flow schemas can be used to represent the processing stages connected by directed information channels. Once the flow model is constructed, names have to be assigned to each of the stages and channels. The main modelling stages are: data provisioning; data collection from external suppliers; data internal creation; data processing (input acceptation, output as well as effects generation); data packaging (storage, aggregation, summarizing, reporting); decision making (human interaction); decision implementation; data delivery; and data consumption.

Knowledge architectures start from communities and their members, and at the end of their journey knowledge transformations occur, measurable in terms of time required for the accomplishment of tasks, cost of operations, resources used and quality of service provided. Since these are measurable outcomes, alternative architectures can be evaluated and selected with respect to their effectiveness or performance, cost efficiency and quality of service. Given the fact that architectures involve users and knowledge processes, which conclude with cognition changes, we can speak about cognitive architectures or cognition-centric systems that include human beings and other cognitive entities with sufficient awareness, learning, language, autonomy and cooperation capability (Mitola, 2009). Cognitive architectures allow such systems to deal with complex operational environments much more efficiently (Sheard and Mostashari, 2009). These architectures also allow a description of the standard problem-solving model, introduced by John Dewey and revised by George Polya, from a distributed intelligence perspective. In this perspective the six stages of the problem-solving process do not appear as constructions of the individual mind but as constructions of thought distributed in the environment, the tools and in other persons, within frameworks of collaboration (Pea, 1993).

3. Activities of innovation

Collaborative innovation is the third and final stage of the roadmap comprising all those activities that allow a community or group of actors to engage in knowledge transformation and offer a better solution to a defined challenge. What really happens in collaborative innovation is that communities and groups of users set in motion knowledge architectures and generate knowledge flow towards new and more efficient solutions. A key instrument here is the concept of 'innovation cycles'.

The idea of innovation cycles, linking theory and practice to cyclical actions that bring understanding, insight and new knowledge, is taken from Action Research (Baskerville, 1999). In particular it comes from user-driven open innovation ecosystems in which situations of openness, cooperation and complex social processes introduce changes into these processes while observing their effects (Schaffers et al., 2008). Thus ideas from users and stakeholders are taken into account when defining the kind of solutions being deployed. Cases of use are identified with respect to users and organizations involved. Prototypes can be tested at early stages allowing iterative modifications of the prototype until they converge towards an accepted and validated solution.

In developing smart environments and ITS applications, a small number of innovation cycles should be implemented. All the cycles have the same structure, with the exception of the first 'preparation cycle' focusing more on the definition of scenarios of use. Following the 'preparation cycle', the next cycles are decreasing in duration due to repetition of the same tasks. As participatory activities evolve, the services deployed converge to a validated solution. Thus main steps of each innovation cycle are:

1. Definition: at this step, based on user feedback carried out at the end of the previous cycle, the modifications to be introduced during the current cycle are defined.
2. Modification and deployment of services: this step is about the implementation of modifications and creation of a new version of the solution/service.
3. Gathering of user feedback in interaction with services: after the deployment of modifications, users should interact with the new version of solution and service.
4. Analysis of user feedback: this step analyses user feedback and draws conclusions about the modifications that were introduced in step 2.
5. Evaluation of the cycle: this is not a step but a milestone. At the end of each cycle an evaluation of the cycle should take place.

With the deployment of innovation cycles, it becomes possible to tap into community and participant creativity, and introduce new ideas and concepts to ITS applications. Brainstorming and scenario building are the methodologies mostly used. Three different types of user contribution to innovation processes can be achieved related to information input, decision-making and solution creation. In this context four different kinds of user can be identified: intuitive, professional, freshman and nerd, based on their familiarity with the application and knowledge objects in question (Reichwald et al., 2004).

5 CONCLUSIONS AND POLICY RECOMMENDATIONS

The evolution of tasks along the roadmap proposed is illustrated by Figure 10.1. The mobilization of CoI enables the selection of actors, the definition of knowledge flows among them and the actors' involvement in innovation cycles, at different time frames of ITS development, implementation and testing. Thus two types of innovations are generated: First, innovations related to new applications and e-services introduced to various transport systems. Innovation at this stage is about the involvement of users in the design and deployment of new e-services. Second, innovations related to the better functioning of transport systems after the

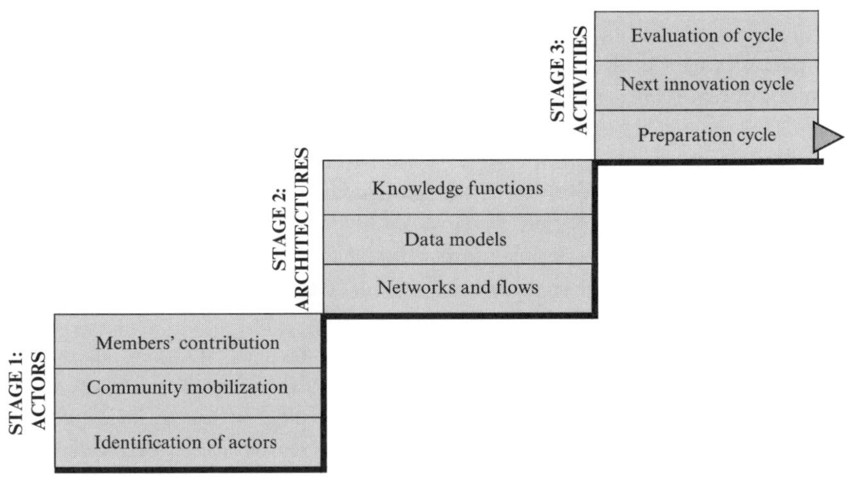

Source: Own elaboration.

Figure 10.1 'Triple A' roadmap stages and steps

introduction of new e-services. Innovation at this stage is about new practices for more efficient networking, co-working, complementary activities and various types of external economies achieved.

The application of the proposed triple A roadmap aiming to improve the innovation chain in the ITS sector presents an array of opportunities and threats as listed below.

The respective opportunities include:

1. Development of successful and sustainable ITS solutions: validated solutions that yield cost savings, have not yet been addressed by the market and have improved market access. After-service support enforces sustainability.
2. Development of a knowledge ecosystem with enhanced innovation capability: capitalizing on the collective intelligence and creativity of CoI enhances the knowledge base and yields new insights and fresh ideas, enabling a multi-perspective confrontation in finding ITS solutions.
3. Rapid uptake and broader acceptance, since ITS solutions reflect user needs and particularities.
4. Problem-solving improvement and rapid technological advancement: bringing CoI into the problem-solving process improves the quality of solutions to transport challenges. The citizens' creative powers, broad feedback and innovation capabilities are harnessed to deliver new and improved services.
5. Flexibility and responsiveness to unpredicted circumstances: risk mitigation becomes more efficient. Broader range and real-time information about the city's ITS systems and operations enable a more agile response to risks and add value to service provision on the side of governments. The aggregation and analysis of vast datasets enables the detection of patterns and anomalies that could not have otherwise been captured.
6. Transparency and trust in government: users view the government as more transparent, accountable and democratic.
7. Sense of equal responsibility: users hold themselves as equally responsible in improving and sustaining ITS solutions.

The respective threats include:

1. Possibility of failure: in action, many efforts to engage CoI in IT solutions development have not produced significant results because of failure to attract users and formulate CoI, poor quality of submitted ideas, ideas conflicting with public policies, insufficient business models and bureaucracy. It is also sometimes difficult to sustain cooperation of multidisciplinary research teams.

2. Not all users might be capable or willing to participate: CoI must be representative of their society mix. However, there is a greater chance that only a fragment of society is represented through them, consisting of technology-savvy users with a desire to influence socioeconomic development.
3. The public sector may appear to be 'inefficient', trying to 'offload' official responsibilities: outsourcing work to CoI may be thought of as a convenient way to discharge its obligations to ITS users.
4. Governance and financing still remain critical: a city's ITS infrastructures unquestionably need to be governed by a regulatory authority and ITS infrastructure development still requires financial investments, which cannot be provided through open innovation procedures.

Taking into consideration all of the above-mentioned aspects, we can assume that public policy, local and regional authorities, in particular, can greatly contribute to the bottom-up design and development of ITS applications and solutions. By endorsing and sustaining CoI and platforms for user-driven experimentation, some metropolitan areas such as Amsterdam, Barcelona and Manchester are becoming transformed into urban living labs that allow business to test and demonstrate innovative solutions for services and utilities (Komninos and Schaffers, 2012). Bottom-up initiatives for open platforms can substantially increase the impact of top-down planning in sustainable transport, energy and environmental excellence.

REFERENCES

Accenture (2011), 'Changing the game: plug-in electric vehicle pilots', available at: http://electricdrive.org/index.php?ht=a/GetDocumentAction/id/27902.

Armstrong, A. and J. Hagel (2000), 'The real value of online communities', in: E.L. Lesser, M.A. Fontaine and J.A. Slusher (eds), *Knowledge and Communities*, Boston: Butterworth-Heinemann.

Bacon, J. (2012), *The Art of Community: Building the New Age of Participation*, 2nd edn, Sebastopol, CA: O'Reilly Media.

Baskerville, R. (1999), 'Investigating information systems with action research', *Communications of the Association for Information Systems*, 2(3), available at http://cais.isworld.org.

Carayannis, E., D. Assimakopoulos and M. Kondo (eds) (2007), *Innovation Networks and Knowledge Clusters: Findings and Insights from the US, EU and Japan*, Basingstoke: Palgrave Macmillan.

European Telecommunications Standards Institute (2012), 'Intelligent transport systems', available at http: //www.etsi.org/index.php/technologies-clusters/technologies/intelligent-transport.

Fischer, G. (2001a), 'Communities of Interest: Learning through the interaction of multiple knowledge systems', Proceedings of the 24th IRIS Conference, Bergen, Norway, pp. 1–14.

Fischer, G. (2001b), 'External and shareable artifacts as opportunities for social creativity in communities of interest', Proceedings of the 5th International Conference on Computational and Cognitive Models of Creative Design, pp. 67–89.

Henri, F. and B. Pudelko (2003), 'Understanding and analysing activity and learning in virtual communities', *Journal of Computer Assisted Learning*, 19, 474–87.

IATA Annual Review (2013), available at http://www.iata.org/about/Documents/iata-annual-review-2013-en.pdf.

Komninos, N. and H. Schaffers (2012), 'Smart cities and the future Internet in Europe', Special Issue, *Journal of the Knowledge Economy*, 2(2).

Komninos, N. (2008), *Intelligent Cities and Globalisation of Innovation Networks*, London: Routledge.

McQuivey, J. (2013), *Digital Disruption: Unleashing the Next Wave of Innovation*, Las Vegas: Amazon Publishing.

Mitola, J. (2009), 'Cognitive radio architecture evolution', *Proceedings of the IEEE*, 97(4), 626–41, available at: http://ieeexplore.ieee.org/xpls/abs_all.jsp?arnumber=4814771.

OECD (2010), *Innovation Strategy – Key Findings*, OECD, available at: http://www.oecd.org/sti/45326349.pdf.

O'Reilly Media (2005), 'What is Web 2.0? Design patterns and business models for the next generation of software', available at: http://oreilly.com/pub/a/web2/archive/what-is-web-20.html?page=1.

Pea, R.D. (1993), 'Practices of distributed intelligence and designs for education', in G. Salomon (ed.), *Distributed Cognitions*, New York: Cambridge University Press, pp. 47–87.

Prat, N., J. Akoka and I. Comyn-Wattiau (2006), 'A UML-based data warehouse design method', *Decision Support Systems*, 42(3), 1449–73.

Radjou, N. (2004), 'Innovation networks: A new market structure will revitalize invention-to-innovation cycles', Innovation Networks Boost Profit series, Forrester, available at: http://sitemaker.umich.edu/mo563/files/innovationnetworks.pdf.

Reichwald, R., S. Seifert, D. Walcher and F. Piller (2004), 'Customers as part of value webs: Towards a framework for webbed customer innovation tools', Proceedings of the 2004 Hawaii International Conference on Computer Sciences (HICSS), Hawaii.

Schaffers, H., J.G. Guzman and C. Merz (2008), 'An action research approach to rural living labs innovation', in P. Cunningham and M. Cunningham (eds), *Collaboration and the Knowledge Economy: Issues, Applications, Case Studies*, Amsterdam: IOS Press, pp. 617–24.

Sheard, S.A. and A. Mostashari (2009), 'Principles of complex systems for systems engineering', *Systems Engineering*, 12, 295–311.

Talluri, K. and G. Van Ryzin (2005), *The Theory and Practice of Revenue Management*, New York: Springer.

T-TRANS (2013a), Deliverable D6.1: Synthesis Report of g-local CIM: Enhancing the transfer of ITS innovations to the market (FP7 research project). Available at: http://www.ttransnetwork.eu/ttrans/synthesis-report-of-glocal-cims/.

T-TRANS (2013b), Deliverable D3.1: ITS State of the Art assessment: Enhancing the transfer of ITS innovations to the market (FP7 research project). Available at: http://www.ttransnetwork.eu/ttrans/its-state-of-the-art-assessment/.

Wenger, E. (1998), *Communities of Practice: Learning, Meaning, and Identity*, Cambridge: Cambridge University Press.

Wenger, E., N. White, J.D. Smith and K. Rowe (2005), 'Technology for communities', in *Guide de Mise en Place et d'animation de Communautés de Pratique Intentionnelles*, Quebec: CEFRIO. Also available at: http://technologyforcommunities.com/CEFRIO_Book_Chapter_v_5.2.pdf.

11. A critical overview of ICT deployment in transport in Europe

Yannis Tyrinopoulos, Anna Kortsari and Maria Chatziathanasiou

1 INTRODUCTION

This chapter aims at providing a critical review and analysis of good practices on the use of information and communication technologies (ICT) in the European transport sector and draws recommendations for the acceleration of the deployment of advanced technologies in Europe. The cases examined are mature industrial applications and products produced by technology providers rather than research results, chosen in an attempt to present practical and applied solutions that stimulate an advanced technological drive in Europe. The chapter serves the aims and goals of this book, as it provides a sound collection of successful ICT implementations, barriers that prohibit the wider ICT deployment, as well as suggestions for tackling these barriers. Therefore, threats and opportunities are directly addressed.

A brief outline of the European transport policy with clear emphasis on ICT for transport is provided below in the current section 1. In this context, the key stakeholders involved in the promotion of ICT in transport in Europe are briefly introduced. The methodology applied for the needs of this chapter is outlined in section 2. The presentation and the analysis of good practices are provided in sections 3 and 4 respectively. The major barriers as well as recommended actions towards the effective deployment of ICT in the European transport sector are discussed in the last section (5).

The application of ICT to the planning and operation of transport systems is widely known as Intelligent Transportation (or Transport) Systems (ITS). There are also several other definitions of ITS, but all have the same meaning involving the use of advanced technologies in the transport sector. The above definition of ITS has been adopted by many countries, and the origin of the term can be attributed to the

United States of America, where it was used to give recognition to the wider application of technology to transit systems, private cars and highways (McQueen and McQueen, 1999, p. 20). At the beginning of the 1990s, the European Commission (EC) used the term 'advanced transport telematics' (ATT) in order to prompt and promote the use of ICT in transport. ATT was the core element in the follow-up to the DRIVE programme within the Telematics Programme (EC Cordis Archives, n.d.) of the EC.

The ultimate role of ITS is to shape future modes of mobility and the transport sector. Through the use of ITS applications, transport can become more efficient, safer and greener (UNECE, n.d.). The various types of ITS rely on radio services for communication and the use of specialized technologies. The most advanced ITS applications can be found in the automotive and aeronautic industries. ITS can include a large variety of technologies, varying from satellite technologies and mobile communications to standards, complex software solutions and Internet-based protocols. Figure 11.1, prepared by ETSI, the European Telecommunications Standards Institute, gives a good illustration of the complexity of ITS

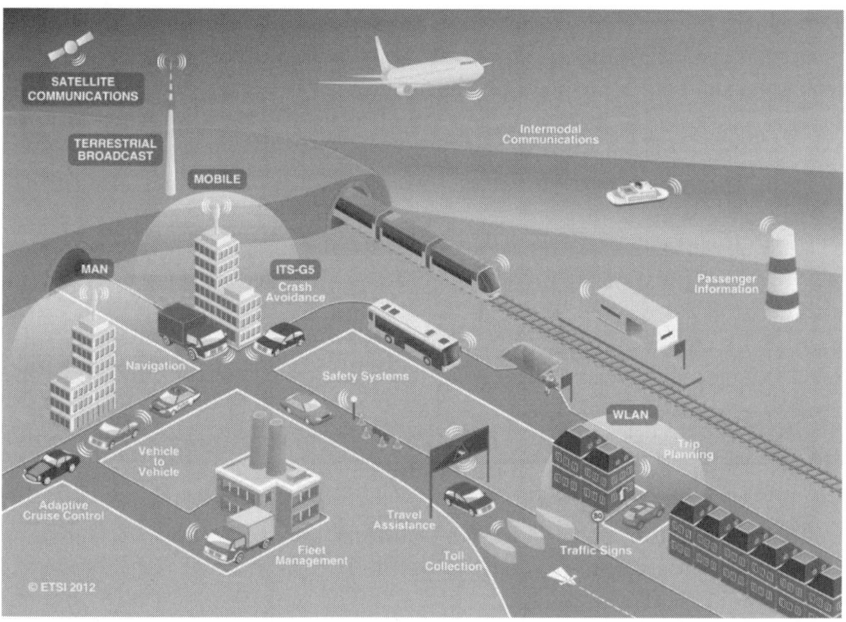

Source: ETSI (2012).

Figure 11.1 The variety of ITS technologies

(ETSI, 2012). The figure depicts the variety of ITS technologies that can be used in the transport sector to support different operations, such as toll collection, crash avoidance, traffic control and many others.

In order to explore the role of ICT in European transport policy, it is necessary, first of all, to briefly review the historical European transport policy and its focus on ITS.

Although the origins of the European transport policy can be found back in the mid-1950s with the Treaty of Rome and the creation of a single market for intra-Community transport, the first serious effort in the development of the Common Transport Policy was made in 1985, when the EC released the white paper on the Completion of the Internal Market (Giorgiand Schmidt, 2002, p. 3). The liberalization of the European transport market was the focal aspect in this document, as a critical prerequisite for regulating the transport sector, primarily through economic instruments. Although liberalization continues to play a central role in the European transport policy, the focus of the 1992 white paper on the Future Development of the Common Transport Policy was shifted to sustainability, primarily environmental, and social cohesion (Giorgiand Schmidt, 2002, p. 3). These two aspects even today retain a high priority in the agenda for European transport policy. A major breakthrough in European transport policy was made in 2001, when the EC published a new white paper, entitled European Transport Policy for 2010: time to decide. Although many of the issues included in this white paper were addressed in the previous one, the new paper was much more comprehensive and complete, and was intended to be a lever, forcing member states to take decisions on outstanding issues. The major challenges identified in this white paper were the imbalance in the development of the different transport modes, the congestion on routes and cities, as well as in airspace and the impact on the environment (EC, 2006, p. 6). Therefore, major emphasis was placed to achieve a significant modal shift away from the road to other, more environmentally friendly, modes of transport, mainly rail (EC, 2001, p. 14).

The first attempts to use ICT in the transport sector took place in the 1980s through the European Parliament's resolution on road safety (1984). This resolution set the basis for the primary developments in informatics and telecommunications, in order to investigate their potential application to road and vehicle safety. At the beginning of the 1990s, the EC started using the term 'telematics', a combination of the terms 'telecommunications' and 'informatics', widely in its policy documents, funding programmes and actions aimed at deploying ICT in the European transport sector. More specifically, the programme 'Telematics Systems of General Interest' was adopted in 1991 (EC Cordis Archives, n.d.). An associated workplan was prepared, in close cooperation with all the major European actors in the field of transport telematics, identifying the requirements for the deployment of transport telematics in Europe.

However, it was only in 2008 that the EC released a dedicated action plan for the deployment of ITS, aiming to promote a more coherent technological drive among the European countries. Through this document, the EC defined specific actions and measures aimed at accelerating and coordinating the deployment of ITS in road transport, including interfaces with other transport modes. The main policy objectives addressed were to make transport and travel cleaner, more efficient, more energy efficient, safer and more secure (EC, 2008, p. 2). In this direction, the EC defined a framework for the coordinated and effective deployment and use of ITS in Europe, which was described in Directive 2010/40/EU (EU, 2010, p. 1). These two documents are currently the most important policy documents for the promotion and deployment of ITS in Europe. They are supplemented with several other regulations targeted at specific ITS fields, such as the harmonized provision of an interoperable EU-wide e-Call (EU, 2013, p. 1).

Figure 11.2 presents a timeline of the key milestones in the European transport policy. The ITS-oriented policy documents are highlighted in bold font.

Today, several European actors and stakeholders are directly or indirectly involved in the promotion of ITS across Europe. Some of them are formally associated with the EC. Below, a few of the most important ones are introduced:

- ERTICO ITS Europe: A network of ITS stakeholders in Europe aiming to accelerate the development and deployment of ITS across Europe and beyond. It was founded in 1991 and today comprises about 100 partners from nine EU member states.
- The European technology platforms bring together stakeholders, led by industry, to define medium to long-term research and technological development objectives and lay down markers for achieving them. Technology platforms play a key role in better aligning EU research priorities to industry's needs. Four technology platforms have been established for all four transport modes (Tyrinopoulos and Chatziathanasiou, 2010, p. 28):
  - ERTRAC (European Road Transport Research Advisory Council)
  - ACARE (Advisory Council for Aeronautics in Europe)
  - ERRAC (European Rail Research Advisory Council)
  - WATERBORNE

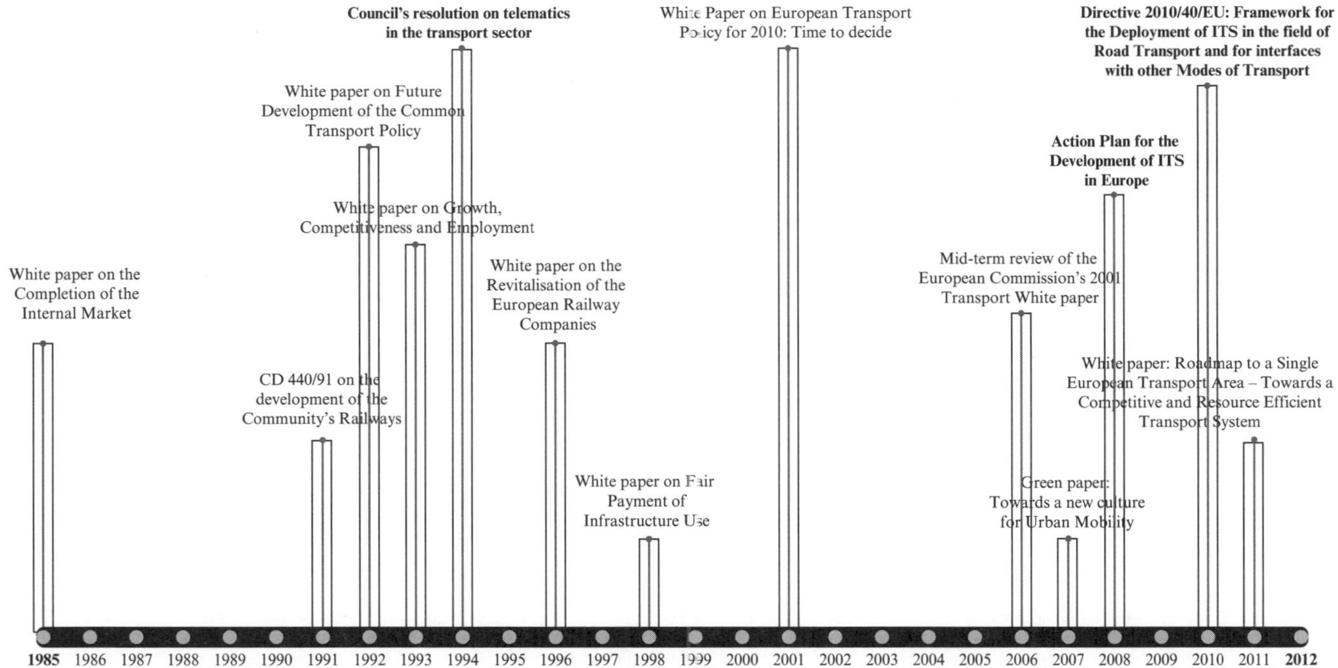

Figure 11.2 Significant milestones in European transport policy development

- EIRAC (European Intermodal Research Advisory Council) has been set up to guide public and private investments in research programmes in the EU, and in the member states or accession countries (EIRAC, n.d.).
- ECTRI (European Conference of Transport Research Institutes) was established in 2003 as a non-profit association with the objective of promoting integrated surface transport research and development in Europe (Tyrinopoulos and Chatziathanasiou, 2010, p. 32).

More information about the above stakeholders can be found through the relevant references of this chapter.

The EC has also set up several working groups and committees on various transport fields, including ITS. The aims of these groups and committees are to examine specific topics related to transport, to provide input to the European transport policy, and to support the implementation of the various action plans released by the EC. Two such groups are the European ITS Advisory Group and the Urban ITS Expert Group. The latter was established in December 2010 and the experts were appointed for a two-year mandate. During its mandate, the group developed guidance on the deployment of three key applications of urban ITS (travel information, traffic management (incl. urban logistics) and smart ticketing), collected together a number of related good practices, and reflected upon the need for further standardization in the domain of urban ITS (EC Mobility and Transport, n.d.).

2 METHODOLOGY

The methodology used in the present chapter includes three major steps: the review of existing ITS good practices, their analysis based on predefined criteria and finally, an overall assessment and provision of recommendations related to the ITS deployment across Europe.

The review of the European ITS good practices was implemented according to the following three levels:

1. Sector: Good practices covering both freight and passenger transport sectors.
2. Mode: Good practices addressing all transport modes and combinations of them (multimodal cases).
3. Thematic area: At this level, the review and reporting was done according to specific thematic areas covering all passenger and freight transport operations across all transport modes. As such, the following

are mentioned: information flow, professional drivers' information services, travellers' information services, electronic payment services, transport of goods, safety and security, resource management and traffic management. Specifically, for road transport, the four priority areas delineated within Directive 2010/40/EU of the European Parliament and of the Council 'on the framework for the deployment of Intelligent Transport Systems in the field of road transport and for interfaces with other modes of transport' were considered, namely:

a. optimal use of road, traffic and travel data;
b. continuity of traffic and freight management ITS services;
c. road safety and security applications;
d. linking vehicles with the transport infrastructure.

Several sources were consulted in order to gather together the good practices and information associated with them, such as EU projects (for example, PROMIT, ENABLE and EasyWay), published information on ITS market products (available on the Web) and information included in the reports of international associations (for example, IATA – the International Air Transport Association).

It is important to underline once more that the focus was placed merely on mature market products and not on research results. This was done in order to exclude applications that are only available as research prototypes and that have only been evaluated in laboratory conditions or small-scale pilot studies. Of course, research products that were eventually exploited by the market have been included. Overviews of European transport research activities (including ITS) are provided by online portals, such as the Transport Research and Innovation Portal (TRIP) (Šitavancová et al., 2010) and the EC Research and innovation portal (EC Research and Innovation Transport, n.d.) and CORDIS archives (EC CORDIS Projects and Results, n.d.).

In total, 70 ITS good practices were identified. This selection reflects an extensive procedure; nonetheless, it cannot be considered exhaustive. ITS applications that have been 'efficiently and effectively' implemented at a very local level but not publicized widely have not been included. Given, also, that the selection was made based on current knowledge, rather 'new' applications that have not yet been evaluated were not examined. For those cases that have been widely promoted to countries all over Europe (for example, the provision of information to drivers through variable message signs (VMSs), ramp metering and so on), only representative country cases are reported based, inter alia, on the available information on their evaluation.

The analysis of the 70 good practices included their mapping against specific criteria defined for that purpose. The selection of the criteria has been driven by the need to inform the target audience about the benefits that can be gained through the use of ITS applications, thus to facilitate their deployment, and to investigate the level of penetration of ICT to the different transport modes, to their operations and to the various European countries. Therefore, the selected criteria refer to the impacts that each ITS implementation has on safety and security, the efficiency of the transport network, the economy, and environment and society; the level of user acceptance; thematic area coverage (as specified above); degree of penetration in the transport modes; and geographical coverage (based on the United Nations' European regions clustering).

It should be noted that the information regarding the good practices' impacts is both quantitative and qualitative. The second approach was necessary due to the fact that, for some ITS applications, no quantitative information was available, despite the fact that the implemented systems are widely known and used by the transport community.

It is also worth mentioning that, to date a vast number of ITS applications have been implemented, the above-mentioned criteria also constituted guidance for the selection of the good practices, meaning that the 70 good practices were selected based on their impacts on, for example, safety and security, efficiency, user acceptance and so on.

The aforementioned analysis concluded in specific results and findings related to ITS deployment in Europe. These findings provided useful information for the third and final step, namely the assessment of the current situation of the ITS deployment in Europe and the provision of recommendations that will stimulate the optimization of the ITS penetration across Europe. Several other sources were also consulted in order to identify existing barriers that hinder the wide deployment of ITS across Europe. These sources included official documents of the EC related to transport policy and ITS, but also research project results, books and-international organizations' reports associated with the subject of this chapter. The authors' experience, through their long presence in EU ITS projects, provided another quite important source.

3 GOOD PRACTICES FOR INTELLIGENT TRANSPORT SYSTEMS

This section aims to provide a general overview of some of the most outstanding European good practices for ITS. According to Ilies et al. (2010), a 'good (best) practice' is a technique, method, process, activity, incentive or reward that is believed to be more efficient and effective at delivering a particular outcome than any other technique, method, process and so on, when applied to a particular condition or circumstance. Good practices can also be defined as the most efficient (least amount of effort) and effective (best results) ways of accomplishing a task, based on repeatable procedures that have proven themselves over time for a large number of people (Gärtner, 2011).

Table 11.1 provides a representative list of 15 (out of the 70) good practices, including at least one good practice for each sector (F: freight or P: passenger) and mode (Rd: road, Rl: rail, A: air, W: waterborne, MM: multimodal). The selection does not imply that these 15 good practices are better than the other 55, rather that they provide a good representation of all modes. The geographical clustering of the good practices was based on the division made by the United Nations: Northern, Western, Eastern and Southern Europe. The table presents brief information about the 15 selected good practices. The aim is to provide a sound indication of the variety of ITS applications in various transport modes and thematic areas across the different European regions. However, the critical review and analysis of the identified ITS good practices presented in section 4 has been carried out for all 70 cases.

4 CRITICAL REVIEW OF ITS GOOD PRACTICES

This section aims to provide a critical review and analysis of the good practices in the field of ITS, which were collected for the needs of this chapter. The criteria used for this analysis are presented in section 2.

All 70 good practices are presented in the Appendix in terms of ID (F: for freight and P: for passenger), name, primary thematic area covered, geographical area/country of application (NE: Northern Europe, WE: Western Europe, EE: Eastern Europe, SE: Southern Europe), transport mode ID (Rd: road, Rl: rail, A: air, W: waterborne, MM: multimodal) and main technological elements used. The Appendix depicts in tabular form the variety of ITS applications reviewed in the framework of this research. These transportation applications have different thematic, technological and geographical origins. Two of these outstanding applications are BRAVO and ERATO from the freight and passenger transport sectors respectively.

BRAVO is a train monitoring and Customer Information System (CIS), which allows the real-time monitoring of cross-border train movements and informs customers of irregularities (Rainer and Sondermann, 2007, pp. 17–18). The system provides the location, relative delay, reasons for delays and statistical analysis of past train runs and also delivers

Table 11.1 Indicative list of good practices in the field of ITS

ID	Name	Short description and main achievements (results and experience)	Thematic area	Geographical area	Mode ID	Main technological elements
F1	Valencia Port Community System (PCS)	The PCS (valenciaportpcs.net) offers a web-platform for optimization of the maritime logistics processes. This is achieved through standardized information exchange, facilitated integration opportunities with other associated ports and sea–port–land integration. 28 shipping companies, 169 logistics operators, 66 maritime agents, 13 empty container depots, and four stevedoring companies are integrated in the platform; more than 90 000 daily messages exchanged; response time five minutes/message. The system integrates all the agents involved in the logistics operations that take place in the port and solves the coordination problems they have to load and unload shipments (Tyrinopoulos and Chatziathanasiou, 2011, Annex I).	Information flow	Southern Europe/ Spain	W	The platform integrates Flat Text Files, XML (extensible Markup Language) and EDI (Electronic Data Interchange) services. It gives users the opportunity either to use web services or to integrate their system to the platform's application (Tyrinopoulos and Chatziathanasiou, 2011, Annex I).
F2	E-TRAIN and GOAL	E-TRAIN is a Global Positioning System (GPS)-based system that enables the real-time management of information flow and train	Information flow	Europe	R1	The e E-TRAIN system uses on-board GPS units and GSM (Global System for Mobile

		monitoring, while GOAL is software that manages transport data in real time. The two systems work together in order to inform customers on their shipments. 98% of the messages exchanged are received from the 'E-TRAIN' system (the remaining 2% is not received due to lack of availability of GPS and network, i.e. inside tunnels and through mountains) (Ruesch et al., 2009, pp. 247–54).				communications) for train monitoring, while GOAL is software developed by Hupac that automatically receives the data from E-TRAIN and delivers information on train movements, whereabouts, arrivals or time variations. XML interfaces are used to acquire and process data that come from peripheral terminals that do not use GOAL (Ruesch et al., 2009, pp. 247–54).
F3	e-freight	The system covers the end-to-end paperless transport process of the air cargo industry, where paper documents are replaced by the exchange of electronic data. The system includes a set of business processes and standards that allow the removal of paper documents from the process of shipping air cargo from origin to destination. Documents are replaced by the exchange of electronic data (IATA, 2013d, pp. 2–3).	Information flow	Europe	A	e-Doc standards used as part of e-freight rely on the use of EDI (Cargo-IMP (Interchange Message Procedures) or XML or scanned images). The system uses the existing air cargo industry messaging infrastructure (in-house technology) (IATA, 2013d, p. 5).

Table 11.1 (continued)

ID	Name	Short description and main achievements (results and experience)	Thematic area	Geographical area	Mode ID	Main technological elements
F4	FREILOT	The system promotes parking toll machines for drivers, which are located in delivery spaces in the centre of Bilbao in order to manage the parking places and reservations. Four FREILOT Delivery Space Booking areas are located in the centre of Bilbao in the areas of Licenciado Poza, Santutxu, General Concha and Pérez Galdós (Zubillaga, 2011, p. 8).	Transport of goods	Southern Europe/ Spain	Rd	The machines are equipped with loop detectors that receive the places occupancy and Programmable Logic Controllers (PLCs) that send the data to the control centre. Vehicles are equipped with chip cards that validate the reservation for the delivery spaces (Flechl, 2013, p. 84).
F5	Dynamic HGV information	This is a system that dynamically informs Heavy Goods Vehicles (HGV) drivers on available free parking spaces and allows pre-booking for private parking sites. Provision of area-wide information on free parking spaces. An Internet service informs drivers and dispatchers on the occupancy rate of parking facilities, as well as when the facility will be fully occupied (EasyWay project brochure, 2009b).	Safety and Security	Western Europe/ Germany	Rd	Information panels for message display; wireless terrestrial magnetic field sensors for counting the vehicles; subtraction model that calculates the number of free parking places (EasyWay project brochure, 2009b).

F6	Ambrogio IT	This is an information technology (IT) system for the management of swapbodies and owned rail wagons (management of customer requests, definition of the operation schedule and production of trains). The system offers standardized information flow within the Ambrogio terminals network, optimization of train composition and better monitoring of the 380 rail wagons and 1300 swap bodies (Ruesch et al., 2009, pp. 255–62).	Information flow	Southern Europe/Italy	MM	Network of computers and software implementation with HTML (Hyper Text Markup Language) pages and SQL (Structured Query Language) database.
F7	CESAR	It is an information portal (Internet communication platform) that provides: harmonization of information exchange; intermodal operator timetable information; booking services; and tracking and tracing between terminals (Ruesch et al., 2009, pp. 42 and 311). The system offers one standardized interface for co-ordinated client communication services. Several operators appear like one virtual company, so that the client works with one single address in order to receive the desirable information about the loading units (http://www.cesar-online.com). Up to 2012, 460	Information flow	Europe	MM	Rail/road operators send status messages to the central system, which forwards them to the customers through the Internet. Virtual client technology is also used.

Table 11.1 (continued)

ID	Name	Short description and main achievements (results and experience)	Thematic area	Geographical area	Mode ID	Main technological elements
F7	CESAR	customer contracts were signed. Each year 9 million status messages are sent and 2.6 million transports are conducted (Burkhardt, 2012).				
P1	Rotterdam Vessel Traffic Services (VTS)	System that provides: information services that support decision-making processes on board vessels (information on position, identity and destination, etc.); traffic organization services that identify and manage potentially dangerous traffic situations; navigation assistance services that support on-board navigation decision-making processes (Röhner, 2013, p. 3). The Rotterdam VTS area extends for 38 nautical miles seawards of the port entrance. The area is divided into sectors, each one having its own dedicated VHF (Very High Frequency) channel.	Traffic management	Western Europe/the Netherlands	W	VTS centres and VTS equipment: Aramis V3000 is used by the traffic centres from January 2012 for the monitoring of the vessel traffic and the identification of individual vessels, tracks, intended routes, etc. The Harbourmaster Management Information System (HaMIS) is also used for the provision of information on each vessel traffic participant (Seignette, 2012, p. 2). Radar technologies, cameras,

						AIS (Automatic Identification System) transponders and other sensors are also used for the creation of a live traffic image.
P2	Trainline's smart card scheme	System that enables all rail ticketing, pay-as-you-go ticketing and customer account management online and on-phone. The system is implemented as a mobile application and web service under the URL: m.thetrainline.com. The web-based platform enables customers to check train times, fares and availability via mobile phone and make a purchase. Tickets are delivered onto customers' smartcards through their PC or to their smartphone directly. Customers can add their ticket to their smartcard at the check-in point at stations or on board when their ticket is checked. They can manage their Smart account on the Internet or via a Java application on their phone (Acklam, 2008).	Electronic payment	Northern Europe/UK	R1	The system includes GMS communication (mobile devices), web services and smartcards.

Table 11.1 (continued)

ID	Name	Short description and main achievements (results and experience)	Thematic area	Geographical area	Mode ID	Main technological elements
P3	PADIS	Passenger and Airport Data Interchange Standards (PADIS) uses: EDI and XML message standards for passenger travel and airport-related passenger service activities. PADIS message standards support the following application areas: publication of schedules and slot management; airline distribution including authenticated shopping; reservations and electronic ticketing; airport resource management and airport handling, including baggage handling; data exchange between airlines and governments concerning passenger data; data exchange between airlines and airports for operational flight-related data; XML schemas for exchange of data on the bar coded boarding pass (IATA, n.d.).	Information flow	Europe	A	EDIFACT (Electronic Data Interchange for Administration, Commerce and Transport), XML schemas.
P4	Meteotrans – European Weather Data Exchange	System for the exchange of meteorological data. It enables the weather data exchange between Germany and the neighbouring countries through an advanced	Information flow	WE/Germany, Luxembourg	Rd	Data exchange is done through the communication protocol DATEX II. The system is connected

		database for operators, thus improving the quality of prediction models (EW EEG/11/7, 2013b, p. 7).				to Germany's National Meteorological Service (DWD). DWD collects data every 15 minutes via public FTP (File Transfer Protocol) access and it provides data every 5 minutes via a second FTP access in BUFR (Binary Universal Form for the Representation of meteorological data) format. Maps are used for visualizing the locations of the measuring points and the measured values of air temperature, surface temperature and temperature at a depth of 5 cm.
P5	CRITER	The Grand Lyon Urban Traffic Management System provides real-time traffic management of traffic lights and real-time information to drivers and travellers. The system enables real-time management of traffic through the management of	Traffic management	Western Europe/France	Rd	Cameras and sensors for traffic monitoring; telecommunication network; software; VMSs for display of information (Grand Lyon, n.d.).

Table 11.1 (continued)

ID	Name	Short description and main achievements (results and experience)	Thematic area	Geographical area	Mode ID	Main technological elements
P5	CRITER	green waves and the provision of traffic light intersection priority to public transport. It links together certain traffic light intersections. This allows prompt detection of malfunctions and prompt repair. Furthermore, drivers and travellers are informed in real-time on traffic status, events, weather forecast, parking places, bike road network, free bike services, car sharing and carpooling. CRITER manages traffic on a 2400 km road network in Lyon. 1096 traffic lights are connected together, 209 video cameras are located on the main road axes of Lyon, 447 measuring stations are used for real-time traffic detections and 42 VMSs (Variable Message Signs) for information display (Grand Lyon, n.d.).				

P6	AnachB.at	This is a traffic information portal providing dynamic information about traffic conditions and intermodal routing service, including a routing service for cyclists and park-and-ride information (Flechl, 2013, p. 15). The platform provides an intermodal network presentation for all traffic modes and an integrated cooperative traffic management combining information, monitoring, controlling and regulation as well as the optimization of the traffic system. Users have free access through website, smartphone app and widget.	Traveller information services	Western Europe/ Austria	Rd	The system is linked with the dynamic traffic data pool, which is supported by numerous ITS partners and the Graph Integration Platform (GIP) that has been developed for the region. GIP is the digital picture of the traffic network of the Vienna Region. The AnachB.at portal is supported, inter alia, by traffic telematics technology (i.e. traffic detection systems that use loop detectors and Floating Car Data collection) and the traffic model of Vienna Region (Catch – MR, n.d.).
P7	Bilrejseplanen.dk	This is a multimodal planner that combines car, bus, train, bicycle, ferries and walking choices. The planner offers: real-time schedules of ferries; historical information about delays on specific travel time for cars; scheduled plans for buses and trains.	Traveller information services	Northern Europe/ Denmark	MM	The planner is fed with data from Rejseplanen public transport and travel time data from the Danish Road Directorate (these are collected via probe

Table 11.1 (continued)

ID	Name	Short description and main achievements (results and experience)	Thematic area	Geographical area	Mode ID	Main technological elements
P7	Bilrejseplanen.dk	All car routes are based on realistic travel times (rush hours are included). The suggested routes are compared to travel time, costs and environmental impact (EasyWay, 2012b, pp. 5–8).				vehicles equipped with GPS).
P8	Traffic Information Agency Bavaria (VIB)	The system is a dynamic journey planner that offers intermodal information on all available transport modes, including pedestrians and cyclists, as well as forecasted road traffic conditions, including weather information (EasyWay project brochure, 2009c). The service is implemented as an Internet platform under the URL: www.bayerninfo.de, and it is also linked to Bavaria's central data hub for public transport (DEFAS BAYERN) (BayerInfo, n.d.).	Traveller information services	Western Europe/ Germany	MM	The service is fed with data from various sources such as: traffic sensors that measure traffic volume and speed of traffic on motorways; reports from Bavarian traffic authorities and operators (i.e. for current incidents, long-term or temporary road works, road closures); Floating Car Data; meteorological stations; traffic cameras (BayerInfo, n.d.).

the estimated time of availability of train paths and terminals. BRAVO enhances efficiency, quality and system technologies in order to achieve a sustainable increase of intermodal transport volumes.

ERATO is a traffic management system that provides traffic management services mainly related to incident management and real-time information to travellers, such as travel times and traffic conditions (EW EEG/11/7, 2010, pp. 9, 16). The information on the ERATO road network is provided to travellers through Variable Message Signs (VMS), and Internet and smartphone applications. The system aims at coordinating the traffic control actions of the two road operators of the urban fast lanes of Toulouse (France) and, thus, reinforcing the coherence and efficiency of the network.

The analysis of the 70 good practices resulted in the diagrams in Figure 11.3, which present the distribution of the ITS applications by

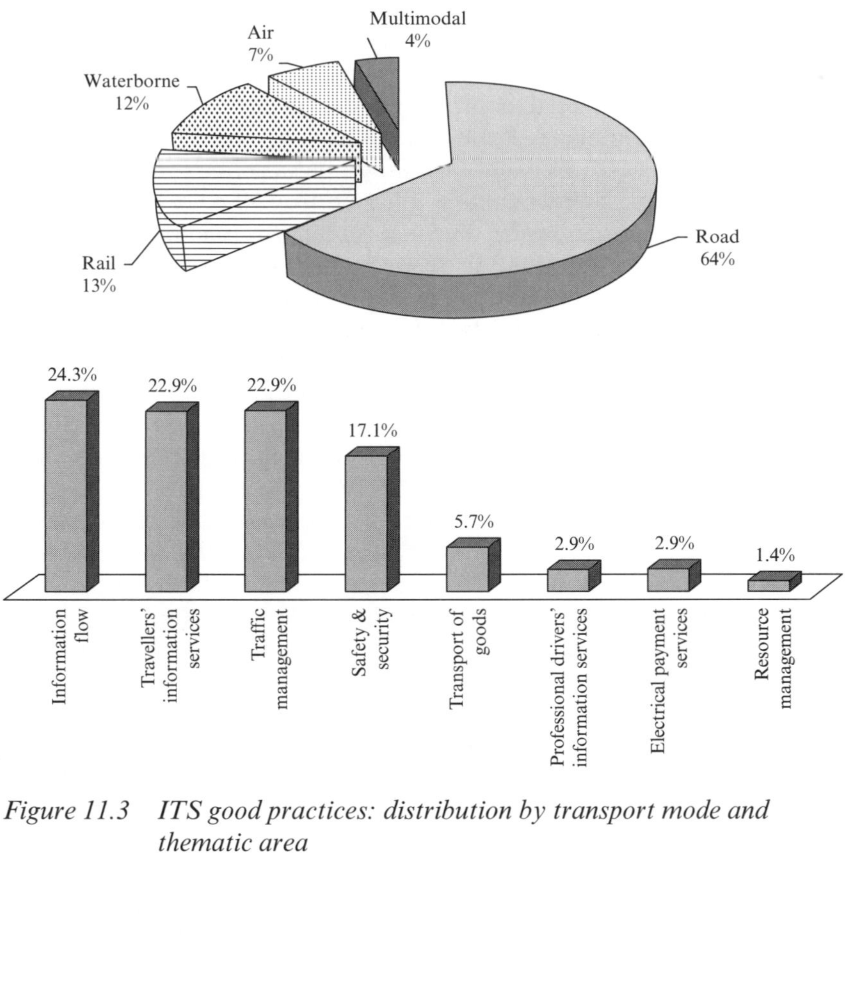

Figure 11.3 ITS good practices: distribution by transport mode and thematic area

transport mode and by primary thematic area addressed. Although the ITS good practices gathered for this analysis comprise just a portion of all the ITS applications available in Europe, the two diagrams are still quite representative of the current situation concerning the level of ITS deployment.

The analysis of the good practices revealed several interesting findings related to ITS deployment across Europe, with the first one being that the ITS industry and business community have concentrated their efforts on road transport, with the vast majority of ITS applications implemented for this transport mode. This can easily be attributed to the predominance of road transport movements: in 2008, passenger cars and buses together accounted for 92.7 per cent of all inland passengers/kilometres, while roads carried 77.5 per cent of inland freighted goods, with intra-EU rail transport in second place at 16.5 per cent (Eurostat Yearbook, 2012, pp. 449–50, 458).

Concerning the road mode of transport, it was found that dynamic traffic management services (such as hard shoulder running, dynamic lane management, ramp metering and so on) have largely been deployed in Western Europe (with the large players located in France and Germany). On the other hand speed control systems, incident warning systems (VMSs) and incident management systems have also been widely deployed in Western Europe (France and Germany), but also in Northern Europe (UK, Denmark and Sweden) and Southern Europe (Spain).

Another activity that has become mainstream for road operators is the delivery of traveller information services. In this respect, traveller and multimodal information services seem to be considered a higher priority than traffic management services (both of which are included in the first priority area of the 2010/40/EU Directive). Finally, driver assistance systems are relatively few and, as expected, implemented mostly in countries with a strong automotive presence (Italy, Germany and France).

Incident management and emergency handling have also been identified as top priority areas for European road ITS systems. This can be easily explained when considering the importance of road safety within the EC agenda (EC, 2010a) and the high societal cost of human fatalities and injuries. Good practices in the field or road safety include both Europe-wide initiatives (e-Call), but also ITS applied in smaller scales (nationally or at a highway/road network level) and they have significant benefits to report (see also impacts below).

When it comes to ITS in inland waterways, this mostly entails river information services (RIS) that support a real-time exchange of data among all the actors involved and are implemented in almost all the European countries that use inland waterway transport (that is, Austria, the Netherlands, Hungary, Romania, Croatia and Serbia) (Gille, 2011, p. 267).

In maritime transport as well, special emphasis was placed on supporting standardized communication procedures (that is, data/document exchange with the use of openly available international standards like UN/EDIFACT or XML technology and paperless procedures).

For air transport, the ITS applications are mostly driven by private initiatives (airline companies); therefore, information on current systems, both for passenger and freight operations is quite limited in terms of its public availability. Thus, the good practices selected here are limited to the initiatives of international organizations, such as the IATA.

Next, information services for freight transport and freight management services have been deployed in the vast majority of European countries, and form the ITS thematic area with the greatest representation across Europe.

In terms of countries, one more finding deriving from the analysis of the good practices is that the Scandinavian countries (Finland, Sweden and Norway), as well as the Netherlands appear to be more advanced in vessel traffic management systems and waterborne ITS applications than the maritime countries of Southern Europe.

On the other hand, countries of Eastern Europe (Bulgaria, Romania, the Czech Republic, Ukraine, Poland and Hungary) still face some important issues that hinder the wider deployment of ITS, such as poor availability of ITS infrastructure, lack of information systems for road users and maintainers, lack of integrated management systems, lack of coordination between institutions and weaknesses in public administration structures, lack of appropriate legislation, restrictions on financing, and lack of national ITS strategies. Similar issues apply in some of the countries of Southern Europe as well (Albania, Slovenia and Croatia) (Kernstock et al., 2013, pp. 50–71). This will be elaborated further in the next section.

From a technological point of view, specific technologies appear to be quite popular in certain transport operations. For example, traffic management systems widely use roadside equipment such as loop detectors, Closed Circuit TVs, dynamic message signs and ramp meters. On the other hand, fleet management systems are strongly supported by technologies that provide the whereabouts of the vehicles (GPS), but also technologies that support the standardized message exchange (such as EDI). More specialized equipment/technology is also used for other operations. This includes, for example: electronic tags for electronic payments; radio communication links and communication protocols that enable the collection of Floating Car Data (for monitoring of the real-time traffic data and the provision of real-time information) and specialized software that supports the operation of Traffic Management Centres.

Table 11.2 presents the impacts of ten indicative good practices. For

Table 11.2 Impacts of indicative ITS good practices

ID	Name	Impacts on safety and security	Impacts on efficiency	Impacts on environment	Socio-economic impacts	User acceptance
F1	BRAVO train monitoring and Customer Information System (CIS)	–	Increase in traffic volumes of about 57% in unaccompanied combined transport on the Brenner corridor (Rainer and Sondermann, 2007, p. 9); 99.9% reliability for transport of documents; maximum delay time 180 min. for trains; 90% punctuality; modal shift: 5.92 to 6.86 million gross tonnes from 2005 to 2006 (EIA, 2010).	–	–	No information available
F2	Southampton Vehicle Booking System (VBS)	–	Reduction in containers waiting times from 4 hours to 30 minutes (Lubulwa et al., 2011, p. 8).	–	–	No information available

F3	ShortSea XML	–	Simplified processes; reduction of errors; more transparency; better vessel utilization; neutral message management (ShortSea XML, n.d.).	–	The system aims at reducing by 10–20% overall costs (i.e. reduced man hours, reduced EDI implementation costs), thus increasing competitiveness (ShortSea XML, n.d.).	High user acceptance (Sundvor, 2007, p. 9).
F4	HGV parking places availability on A10 Motorway	–	Occupancy of the Val Neuvy parking lot increased by more than 25% (EW EEG/11/7, 2012b, p. 14).	–	–	Nearly half of the drivers use this information in order to choose the place where they will stop, while 95% find the messages comprehensible and useful (EW EEG/11/7, 2012b, p. 13).

Table 11.2 (continued)

ID	Name	Impacts on safety and security	Impacts on efficiency	Impacts on environment	Socio-economic impacts	User acceptance
F5	Automatic Number Plate Recognition (ANPR) system for heavy vehicle control	The number of people involved in accidents has been reduced after the system's implementation, as well as the severity of injuries (EW EEG/11/7, 2012a, pp. 15–16).	Decrease in the road density during the average hour and in the maximum flow rate hour. Improvement in the Level of Service (LOS) – segments with LOS F are less than before the system implementation, while sections with LOS D and E have increased considerably (EasyWay EEG/11/7, 2012a, pp. 18–20).	400 euros/day savings due to reduction of CO_2 emissions. Noise reduction has also been expressed as a 2 353 536 euro saving (EWEEG/11/7, 2012a, p. 22).	–	No information available
P1	Dynamic Speed Limits	–	Decrease of 7% in travel times, which was beneficial for 39% of the road users; increase of 8% in road capacity due to the increase in the speed limit from	Reduction in the number of days exceeding the concentration norm of PM10 by 1.9 days – from 24.4 to 22.5	–	High user acceptance (Van der Veen and Taale, 2011, p. 20).

			80 to 100 km/h at the start and the end of the evening peak period; decrease by 200–400 per day of the average number of vehicle hours lost due to delay (Van der Veen and Taale, 2011, pp. 6–7).	days (Van der Veen and Taale, 2011, pp. 6–7).		
P2	Traffic Management System in highway sections	Reduction of approximately 18% of traffic accidents, 10% of traffic accidents with injured people, 11% of injured people and 18% of persons with heavy injuries (Merlin and Vindišar, 2009, p. 23).	–	–	–	80% of the drivers were positive about the usefulness of the system, its contribution to the reduction of traffic accidents and its comprehensiveness (Merlin and Vindišar, 2009, p. 27).

Table 11.2 (continued)

ID	Name	Impacts on safety and security	Impacts on efficiency	Impacts on environment	Socio-economic impacts	User acceptance
P3	Traffic Scotland Web Information Services	–	Improved decision making due to the provided information leading to greater network efficiency (Watson et al., 2008, p. 10).	Environmental gains due to a 6% modal shift to public transport – if extrapolated to total users this means up to 10000 people using public transport more often (Watson et al., 2008, p. 11).	–	67% of survey respondents check the website daily or at least twice a week; 82% of the respondents have changed their journey plans due to the information provided; 25% have made long-term changes to their travel plans and 6% have been persuaded to take public transport more often (Watson et al., 2008, p. 4).
P4	Real Time Public Transport Information (RPTI)	–	Reduced average passenger waiting times on bus stops (EW EEG, 2013d, p. 10).	–	Net socio-economic benefit of 18:1 (EW EEG, 2013d, p. 10).	Increased user satisfaction – 75% of users were satisfied with the service (EW EEG, 2013d, p. 6).

P5	ERATO traffic management system	Decrease of 11% of the accidents during the period 2005–07 (versus the period 1998–2004) (EW EEG/11/7, 2010, p. 6).	63% of the drivers change their route (if indicated that an incident occurred), among which 25% do it systematically (EW EEG/11/7, 2010, p. 17).	Given that ERATO imposed a speed limit reduction from 110 km/h to 90 km/h, between 2004 and 2008 the following have been measured: 13.3% reduction in unleaded consumption, 0.2% reduction in diesel consumption, 46.5% reduction in CO_2 emissions, 23.2 reduction in NOX, 35.4% reduction in COV, 3% reduction in CO_2 and 85.2% reduction in SO_2 (EW EEG/11/7, 2010, p. 6).	–	High user acceptance (especially where the information via VMS is concerned) (EW EEG/11/7, 2010, p. 16).

the purposes of the current section, emphasis was placed on those good practices that produce quantitative effects. The ITS good practices data collection process demonstrated the actual role of ITS in the transport sector, as mentioned in the introductory section, that is, through the use of ITS applications transport will become more efficient, safer and greener (UNECE, n.d.). In certain cases, the tangible benefits, associated with clear figures, offer a clear message: ITS is a powerful tool in the hands of policy makers and industrial players that can be used to generate environmental and energy gains, improve safety and security, and pave the way for economic growth.

The ITS good practices investigated here reveal substantial benefits for all target groups of the transport sector, such as freight operators, transport network managers and travellers. In addition to the quantified impacts presented in Table 11.2, the ITS applications are capable of generating benefits that more or less apply to the majority of cases. In this respect, important benefits for freight operators and transport network managers include: better use of existing infrastructure, reduced operational costs, security of data exchange and increased capacity of the transport network. For passengers and travellers, significant benefits have also been identified. Some of them are: reduced congestion and emissions; reduced travel times/waiting times; and increased quality of public transport services.

In order to provide some specific examples, important benefits are reported for specific areas of ITS such as incident and emergency handling/management in road transport and electronic data interchange in maritime logistics. For incident management these include: improvement of drivers' safety, limitation of the risk of accidents, of collateral damages and of secondary incidents (leading to reduced societal costs), improved traffic fluidity and improved information provided to the drivers. Some interesting quantitative results are presented in evaluation of the e-Call in Finland (Virtanen et al., 2006, p. 16). The evaluation has indicated that the e-Call system would have very likely prevented 3.6 per cent of all traffic fatalities in years 2001–03. Important benefits for road safety were also reported by traffic management systems, like the VMS deployment on the Slovenian highways, where a reduction of approximately 18 per cent of traffic accidents, 10 per cent of traffic accidents with injured people, 11 per cent of injured people and 18 per cent of persons with heavy injuries had been reported (Merlin and Vindišar, 2009, p. 23). Improving road safety is a key result for many ITS good practices, despite the fact that in many cases it is anticipated as a 'rebound' effect, that is, thanks to the improvement of general flow conditions and the stabilization of the average speed (SINA, 2009).

For electronic data/document exchange in maritime transport, important benefits have also been generated, such as: reduction of processing costs, reduction of transaction time and increase in the accuracy and security of data exchange (real-time visibility of the exchange), thus resulting in improved business efficiency (savings on human resources, better customer service, reduced cancelled orders). It is interesting to notice, as a quantitative example, that the ShortSea XML open message standards can lead to a reduction of 10 to 20 per cent of the overall administration costs of shortsea shipping (ShortSea XML, n.d.).

5 DISCUSSION, CONCLUSIONS AND RECOMMENDATIONS

The effective deployment of ITS solutions across European countries is a major challenge, clearly depicted in the many EU policy documents on the subject, primarily the ITS Action Plan (EC, 2008, p. 3) and Directive 2010/40/EU (EU, 2010, p. 1). In this respect, drawing up recommendations that will facilitate and accelerate this process is of vital importance. Before doing that, it is essential to shed some light on the current situation and the major barriers that prevent or discourage not only the wide deployment of ITS in Europe, but also its penetration in the transport sector. Some of these barriers were also identified through the analysis of good practices presented in the previous section, while some others have been derived from the long involvement of the authors in the implementation and deployment of ITS applications in Europe. The identification of barriers/threats and the provision of suggestions/opportunities for the wider ITS deployment across Europe directly address the aims of this book.

Through Directive 2010/40/EU, the EC has established a framework for the coordinated and effective deployment and use of ITS in the EU. According to this framework, the current status of ITS deployment in Europe is characterized by Ferreira (2010, p. 7): fast technical development that leads to a high number of mature applications; fragmented and uncoordinated deployment; little continuity of services and a low degree of intermodality. This framework has also identified some of the major barriers that currently prevent the wide deployment of ITS in the European transport sector (Ferreira, 2010, p. 7): lack of interoperability between systems and between services; lack of effective cooperation among stakeholders and/or member states and privacy and liability yet to be clarified.

In one of its publications, the United Nations Economic Commission for Europe (UNECE) collected and analysed the various gaps and stumbling blocks in the progress towards ITS deployment (UNECE, 2012, p. 87).

These included, among others, the limited political will regarding ITS, the lack of harmonized policies among the UNECE member states, the different speeds of progress in the public and private sectors, the lack of or limited ITS training, and the fragmentation of technical standards. Even the lack of a commonly agreed definition for ITS has been identified as an obstacle.

Further to the above, additional barriers include: unbalanced deployment of ITS among EU member states, transport modes and major transport players and SMEs; lack of or limited funding that would facilitate investments; reluctance from some transport operators (especially SMEs) to adopt advanced technologies (either due to financial restrictions or a lack of specific quantified benefits); bureaucratic obstacles and a lack of legislation in certain European countries; lack of or limited cooperation between the research/academia and industrial sectors; lack of quantified/tangible benefits from the use of mature ITS applications that would prompt other transport stakeholders to invest in similar solutions; and low visibility of outstanding ITS applications.

Barriers to the wide deployment of ITS are also imposed by obstacles to transport innovation. Wiesenthal et al. made, on behalf of the EC Joint Research Centre, an interesting assessment related to innovation in the European transport sector. Their report presents a long list of barriers to transport innovation. Some of the most important are (Wiesenthal et al., 2011, p. 42): established enterprises that dominate markets and therefore make it difficult for newcomers to enter; complex innovation systems require coordinated innovation efforts between all players, including industry, academia and governments; lack of information on the benefits of innovation and uncertainty about potential regulatory changes create uncertainty in market demand and therefore little incentive to innovate; a lack of qualified personnel ('human capital'); and problems with financing.

Directive 2010/40/EU prioritizes certain actions aimed at accelerating ITS deployment in the EU. This entails the following four action areas (EU, 2010, p. 4), which are mainly oriented towards road transport:

1. optimal use of road, traffic and travel data;
2. continuity of traffic and freight management ITS services;
3. ITS road safety and security applications;
4. linking vehicles with the transport infrastructure.

The above four priority actions can be seen as focal transport topics, upon which the EC is concentrating its ITS deployment efforts. However, the barriers and obstacles presented above together create a complicated environment that discourages the wide deployment of ITS in Europe.

Concrete and specific actions need to be taken at several levels, such as technological, operational and legislative, in order to stimulate the fast uptake of ITS applications and coordinated ITS deployment across the EU.

The barriers and obstacles themselves point the way regarding the actions that can be taken towards wide implementation and uptake of ITS in Europe. This goal can be achieved through several recommended actions, with the first of them being the promotion of standardization and interoperability. This will be essential in order to avoid the emergence of a patchwork of ITS applications and services (EC, 2010b, p. 9). The interoperability of systems, achieved by adopting widely acceptable standards represents a crucial success factor for the wider adoption of ITS. The active participation of major industrial players in the creation of standards is essential. Furthermore, particular attention should be paid to efforts that will accelerate the formation of regulations and agreements across countries on technical compatibility and interoperability, which will contribute to the continuity of services across borders.

Furthermore, it is necessary to promote legislation and harmonized policies. In many European countries, legislation is a strong barrier to the use and wider promotion of ITS. The existence of a solid legislative framework and harmonized national policies would facilitate the effective planning and implementation of ITS solutions by the private sector, and enable reliable and safe services to be offered to hauliers and passengers.

The boost of technological integration is also necessary. As Sussman stated, the fundamental strength of ITS is also the most important barrier to its deployment (2005, p. 82). The fundamental insight of ITS is that the infrastructure and the vehicles that operate on it perform as a system. A variety of technologies, from communications, to computers, to sensors and so on, need to be integrated effectively.

The creation of human capital is of vital importance. Skilled personnel are particularly important for strengthening companies' capacity to use advanced ITS solutions, and also to foster innovation. Thus, the certification and education/training of staff is a strong factor, not only for increasing the competitiveness of the European transport sector, but also for facilitating the deployment of ITS applications.

Furthermore, it is necessary to build a consensus among various stakeholders and allocate clear roles to them. Many of the different transport stakeholders have varying perspectives regarding the necessity and implementation of ITS. Often, they have different priorities and even conflicting interests. Therefore, it is imperative to establish a productive dialogue among the key stakeholders, involving all types of actors (industry, academia and governmental agencies) and to define clear priorities for ITS deployment.

A useful tool in this regard could be the use of an ITS architecture, which by default would consolidate priorities, challenges and ITS services. The ITS architecture is a framework defining the technical, institutional/organizational, and commercial features of the future system in an outline and graphical format, showing how all the individual subsystems and components work together (McQueen and McQueen, 1999, p. 59). ITS architecture should be seen as a planning tool and its benefits need to be better understood (UNECE, 2012, p. 92). In Europe, some countries have established their own ITS architectures, while in many others the term is totally unknown in governmental agencies.

In many ITS solutions that cover a particular transport community (such as the port community, for example), the allocation of roles and responsibilities among the members of the community that will cooperate in the deployment of an IT application is a critical success factor. In the case of the Valencia Port Community System, the Port Authority took the role of champion, guiding the rest of the players of the port community in a fruitful cooperation.

The generation of awareness of successful ITS cases based on quantified benefits should also be aimed at. ITS good practices and their benefits should be widely promoted. This needs to be better organized. Currently, there are too many reports and publications on good practices, with very limited information on the quantifiable gains. However, and above all, there should be a quantified analysis (for example, a cost–benefit analysis) of the expected gains from the adoption of successful practices. In most cases, investment decisions regarding ITS are based on the awareness and understanding of possible options, steered by perceptions of benefits versus costs. Therefore, the availability of such data would provide evidence of success and encourage other transport operators and businesses to follow, especially SMEs. In freight transport, for example, the large industrial players dominate technology adoption. However, the entire sector cannot move forward without the technological upgrading of SMEs. As a result, there is a strong need for lowering the barriers-to-entry for SMEs with respect to the utilization of ITS (Ruesch et al., 2009, p. 354). This can be done through various actions, among them the promotion of good practices, accompanied by their quantified benefits.

In addition to the above, many other recommendations could be mentioned, such as boosting innovation, addressing liability issues and enhancing public–private partnerships. No matter where one places the focus, it has to be remembered that the successful deployment of ITS is a complicated goal and a major challenge. Although ITS has a systemic nature, it has other implications too, related to policy, finance, legislation, education and other aspects. These elements need to be addressed effectively and in a synergetic manner in order to enable the fast uptake of ITS, the generation of scale effects leading to cost reductions for transport operators, the promotion of the European ITS industry and the achievement of the ultimate goal of the European transport policy: safe, clean and efficient transport.

BIBLIOGRAPHY

AbOvo (n.d.), *Rail Cargo System – Product Description*, available at http://www.ab-ovo.nl/sites/default/files/corporate/casestudies/RCS-Productdescription 2.0.pdf (accessed 4 September 2013).

Acklam, Jeremy (2008), *Innovative Smart Ticketing in the rail industry*, available at http://www.moving-on-conference.co.uk/pdf/presentations/Tues/morning/JeremyAcklam.pdf (accessed 4 October 2013).

Aulicino, Angelo and Fabrizio Fioravanti (2006), *ICT in Dry Port (Freight Village) Management*, available at http://www.port-net.net/activities/pdf/aulicino_fioravanti_alessandria_12102006.pdf (accessed 22 August 2013).

BayernInfo (n.d.), available at http://www.bayerninfo.de/news/bayerninfonews (accessed 4 October 2013).

Bressler, Andrea and Guido Schuster (2008), *Telematics-Controlled Truck Parking at the Motorway A3 Service and Rest Area 'Montabaur'*, evaluation report of the EU-funded project EasyWay.

Brown, Rhona and Linda O'Connor (2004), *NADICS Journey Time Planner – STREETWISE Evaluation Results Report*, TEMPO Secretariat.

Burkhardt, Martin (2012), *Co-operative European System for Advanced Information Redistribution*, available at http://www.unece.org/fileadmin/DAM/trans/doc/2012/wp24/ECE-TRANS-WP24-2012-Pres13e.pdf (accessed 2 September 2013).

Catch – MR (Metropolitan Regions) (n.d.), *Intermodal traveller information – AnachB.at*, available at http://www.catch-mr.eu/public/DB_Data/files/Downloads/CMR_Intermodal_traveller_information_Vienna.pdf (accessed 19 September 2013).

Cohen, Simon, Maurice Aron and Regine Seidowsky (2010), 'A cost–benefit assessment of a dynamic managed lanes operation', *paper presented at the 12th World Conference on Transport Research Society*, Lisbon.

Dörge, Lone (2009), *Multi-modal Traveller Information Service*, Trafikken.dk/Hovedstaden, Denmark, project EasyWay.

Dölger, Reiner (2011), *Mobility Portal Rheinland-Pfalz*, project EasyWay.

DoRIS (n.d.), 'System DoRIS', available at http://www.doris.bmvit.gv.at/en/system_doris/ (accessed 8 October 2013).

EasyWay (2009), *Project Evaluation Summary – Geolocation of Service Vehicles*, evaluation report of the EU-funded project EasyWay.

EasyWay (2012), *Evaluation of Bilrejseplanen.dk*, evaluation report of the EU-funded project EasyWay.

EasyWay (n.d.), *Reyers Tunnel, Brussels*, evaluation report of the EU-funded project EasyWay.

EasyWay Evaluation Expert Group (EW EEG) (2011), *Evaluation of Regional Traffic Information Service*, evaluation report of the EU-funded project EasyWay.
EasyWay Evaluation Expert Group (EW EEG) (2013), *Evaluation of Public Transport Measures in Trondheim – Norway*, evaluation report of the EU-funded project EasyWay.
EasyWay Evaluation Expert Group 11/7 (EW EEG/11/7) (2010), *ERATO Traffic Management System – Real Time Information Services through VMS on Urban Fast Lanes of Toulouse*, evaluation report of the EU-funded project EasyWay.
EasyWay Evaluation Expert Group 11/7 (EW EEG/11/7) (2011), *All Types of Equipment for Road Monitoring. Management and Operation TCC/TICs. Madrid*, evaluation report of the EU-funded project EasyWay.
EasyWay Evaluation Expert Group 11/7 (EW EEG/11/7) (2012a), *ANPR System to Control Heavy Vehicles in the A-8 near Bilbao*, evaluation report of the EU-funded project EasyWay.
EasyWay Evaluation Expert Group 11/7 (EW EEG/11/7) (2012b), *HGV Parking Places Availability Service on COFIROUTE Network*, evaluation report of the EU-funded project EasyWay.
EasyWay Evaluation Expert Group 11/7 (EW EEG/11/7) (2013a), *Development and Implementation of a System of Remote Traffic Lights*, evaluation report of the EU-funded project EasyWay.
EasyWay Evaluation Expert Group 11/7 (EW EEG/11/7) (2013b), *Meteotrans – European Weather Data Exchange*, evaluation report of the EU-funded project EasyWay.
EasyWay Evaluation Expert Group 11/7 (EW EEG/11/7) (2013c), *Highways Agency Weather Information Service*, evaluation report of the EU-funded project EasyWay.
EasyWay Evaluation Expert Group 11/7 (EW EEG/11/7) (2013d), *Traffic Control and Management System in AEBT – BRISA Group*, evaluation report of the EU-funded project EasyWay.
EasyWay Evaluation Expert Group STR 36 (EW EEG/STR 36) (2013), *Traffic Wales Website – Freight Management*, evaluation report of the EU-funded project EasyWay.
EasyWay project brochure (2009a), *Integration of Traffic Information and Google Maps*, available at http://www.its-toolkit.eu/2decide/toolkit/files/ITS_TOOLKIT_333_1298914595.pdf (accessed 2 October 2013).
EasyWay project brochure (2009b), *Dynamic HGV Information at A5 in Baden-Wuerttemberg*, available at http://www.its-toolkit.eu/2decide/toolkit/files/ITS_TOOLKIT_76_1322754404.pdf (accessed 2 October 2013).
EasyWay project brochure (2009c), *Traffic Information Agency Bavaria (VIB)*, available at http://www.its-toolkit.eu/2decide/toolkit/files/ITS_TOOLKIT_9_1298815763.pdf (accessed 2 October 2013).
EasyWay project brochure (2009d), *French–Spanish Cross Border Traffic Management and Information Interoperability of Services*, available at http://www.its-toolkit.eu/2decide/toolkit/files/ITS_TOOLKIT_23_1298819235.pdf (accessed 1 October 2013).
EC CORDIS Archives (n.d.), 'History of the Deployment of transport Telematics', available at http://cordis.europa.eu/telematics/tap_transport/intro/benefits/history.htm (accessed 2 October 2013).

EC CORDIS Projects and Results (n.d.), available at http://cordis.europa.eu/projects/home_en.html (accessed 20 January 2014).

EC Research & Innovation Transport (n.d.), available at http://ec.europa.eu/research/transport/projects/index_en.htm (accessed 20 January 2014).

EIRAC (n.d.), 'About EIRAC', available at http://www.eirac.eu/ (accessed 30 September 2013).

ERTICO ITS Europe (n.d.), 'About ERTICO', available at http://www.ertico.com/about-ertico/ (accessed 30 September 2013).

European Commission (EC) (2001), *European Transport Policy for 2010: time to decide*, European Communities Publications, Luxemburg.

European Commission (EC) (2006), *Keep Europe moving – Sustainable mobility for our continent – Mid-term review of the European Commission's 2001 Transport White Paper*, Communication from the Commission (314), Brussels.

European Commission (EC) (2008), *Action Plan for the Deployment of Intelligent Transport Systems in Europe*, Communication from the Commission (886), Brussels.

European Commission (EC) (2010a), *Towards a European road safety area: policy orientations on road safety 2011–2020*, Communication from the Commission (389), Brussels.

European Commission (EC) (2010b), *Intelligent Transport Systems: EU-funded research for efficient, clean and safe road transport*, Directorate-General for Research.

European Commission (EC), Mobility and Transport (n.d.), *Intelligent Transport Systems for Urban Areas*, available at http://ec.europa.eu/transport/themes/its/road/action_plan/its_for_urban_areas_en.htm (accessed 8 October 2013).

European Intermodal Association (EIA) (2010), *Intermodal Freight Transport & Logistics Best Practices*, available at http://www.eia-ngo.com/wp-content/uploads/2010/01/Best-Practice_Promit.pdf (accessed 3 September 2013).

EU-SPIRIT available at http://www.eu-spirit.com/ (accessed 18 September 2013).

European Telecommunications Standards Institute (ETSI) (2012), *Intelligent Transport Systems*, available at http://www.etsi.org/technologies-clusters/technologies/intelligent-transport (accessed 11 October 2013).

European Union (EU) (2010), *Directive 2010/40/EU of the European Parliament and of the Council on the framework for the deployment of Intelligent Transport Systems in the field of road transport and for interfaces with other modes of transport*, Legislative Act, Official Journal of the European Union.

European Union (EU) (2013), *Commission Delegated Regulation (EU) No 305/2013 supplementing Directive 2010/40/EU of the European Parliament and of the Council with regard to the harmonised provision for an interoperable EU-wide eCall*, Regulation, Official Journal of the European Union.

Eurostat Yearbook (2012), *Transport*, available at http://epp.eurostat.ec.europa.eu/cache/ITY_OFFPUB/CH_10_2012/EN/CH_10_2012-EN.PDF (accessed 15 October 2013).

Ferreira, Francisco (2010), *Directive 2010/40/EU: Framework for the Deployment of Intelligent Transport Systems*, European Commission, Directorate General Information Society.

Flechl, Barbara (2013), *Deliverable D3.2: Mapping of European best practices in ITS*, project SEE-ITS, available at http://www.seeits.eu/docs/Publications/Deliverables/SEEITS-D3.2%20Mapping%20of%20European%20best%20practices%20in%20ITS.pdf (accessed 18 September 2013).

Fortum, Astrid (n.d.), *Road Tolling in Norway – a brief introduction*, Norwegian Public Roads Administration (NPRA), available at http://www.innanrikis-raduneyti.is/media/frettir/Fyrirlestur_Astrid_Fortun.pdf (accessed 3 October 2013).
Gärtner, Markus (2011), 'Best practices for "Best practices"', available at http://www.shino.de/2011/03/21/best-practices-for-best-practices/ (accessed 10 October 2013).
Gille, Johan (2011), *European Good Practices Report for Inland Waterway Transport*, project Platina.
Giorgi, Liana and Michael Schmidt (2002), 'European Transport Policy – A Historical and Forward Looking Perspective', *German Policy Studies*, **2** (4).
Grand Lyon (n.d.), available at http://www.grandlyon.com/ (accessed 5 September 2013).
Hirou, Catherine and Alexia Journe (2005), *Totem Evaluation Report*, TEMPO Evaluation Expert Group, Centrico.
IATA (2010a), *Simplifying the Business Matchmaker*, available at http://www.iata.org/whatwedo/stb/Documents/stb-matchmaker-factsheet-final.pdf (accessed 4 October 2013).
IATA (2010b), *Simplifying the Business, Fast Travel Vertical Campaign for CUSS Airports: Bag Tag printing, StB Phase 11*, available at http://www.iata.org/whatwedo/stb/fast-travel/Documents/Fast-Travel-Bag-Tag-Campaign-Report-July-2010.pdf (accessed 4 October 2013).
IATA (2013a), *e-CSD instructions*, available at http://www.iata.org/whatwedo/cargo/security/Documents/iata-consignment-security-declaration.pdf (accessed 3 October 2013).
IATA (2013b), *e-AWB factsheet*, available at http://www.iata.org/whatwedo/cargo/e/eawb/Documents/e-awb-factsheet-2013.pdf (accessed 4 October 2013).
IATA (2013c), *e-AWB penetration report*, available at http://www.iata.org/whatwedo/cargo/e/eawb/Documents/e-awb-monthly-report-r17.pdf (accessed 4 October 2013).
IATA (2013d), *e-freight fundamentals*, available at http://www.iata.org/whatwedo/cargo/e/efreight/Documents/e-freight-fundamentals.pdf (accessed 3 October 2013).
IATA (n.d.), available at http://www.iata.org/whatwedo/workgroups/Pages/padis.aspx (accessed 4 October 2013).
Ilies, Liviu, Emil Crisan and Ioana Natalia Muresan (2010), 'Best Practices in Project Management', *Review of International Comparative Management*, **11** (1), 43–51.
Kernstock, Wolfgang, Mihai Niculescu and Cosmin Bejan (2013), *Complete consolidated report on the status of current ITS deployment in SEE*, Deliverable D3.1.2 of the EU project SEE-ITS.
Lubulwa, Godfrey, Adam Malarz and Shun Peng Wang (2011), *An investigation of best practice landside efficiency at Australian container ports*, Australasian Transport Research Forum 2011 Proceedings, Adelaide, Australia.
McQueen, Bob and Judy McQueen (1999), *Intelligent Transportation Systems Architectures*, Boston and London: Artech House.
Merlin, Bojan and Borut Vindišar (2009), *Evaluation of VMS deployment on Slovenian part of corridor V*, European Commission – DG-TREN – TEMPO Secretariat.
Mitsakis, Evangelos and Jose Maria Salanova (2011), 'Intelligent urban mobility management and traffic control system for the improvement of the urban

environment quality in the central area of Thessaloniki's agglomeration', *ITS Europe Conference*, Lyon, France.
Permala, Antti, Jarkko Lehtinen and Karri Rantasila (2010), *Best Practices Handbook*, Deliverable D2.1 of the EU-funded project ENABLE.
Prevoisin, Damien and Jerome Ferre (2006), *Automatic Incident Detection and Fast Alert Evaluation Report*, TEMPO Evaluation Expert Group, Centrico.
Rainer, Mertel and Klaus-Uwe Sondermann (2007), *BRAVO Final Report for Publication*, project BRAVO.
Revill, Steven, Joanne Casey, Jonathan Moran and Ian Cater (2009), *Own Language Information System*, Transport Scotland for the EasyWay Secretariat and the Evaluation Experts Group.
Riley, Paul and Alexandra Holubova (2008), *Ex-post evaluation of ITS applications implemented on highway D8 in Czech republic*, Jacobs Consultancy spol.sr.o.
Rittscher, Jens, Semir Husagic, Armin Müller and Matthias Gropp (2008), *State of the Art: Information and Communication Technology: Objectives*, Deliverable D2.2 of the EU-funded project RETRACK.
Röhner, B. (2013), *VTS Communication Procedure*, Port of Rotterdam Authority, available at http://www.portofrotterdam.com/en/Shipping/communication-vts/Documents/VTS-Communicatie-Procedure-2013.pdf (accessed 11 October 2013).
Ruesch, Martin, Cornelia Petz, Urban Kaiser, Constantinos Antoniou, Alexis Gateley and Iakovov Dallas (2009), *Best Practice Year 1, 2 and 3*, Deliverables D3.1–D3.3 of the EU-funded project PROMIT.
Schagala, Roman, Tanja Weber, Martin Müllner and Martin Nemec (2011), *TMC plus service implementation in Austria – Evaluation report*, evaluation report of the EU-funded project EasyWay.
Segura, Jaume and Ramón Cirilo (2005), *ARENA (Accidentes de tráfico: Recogida de iNformación y Análisis): a system for reporting road accidents*, TEMPO Evaluation.
Seignette, Raymond (2012), *Vessel traffic management in the Port of Rotterdam*, available at http://www.porttechnology.org/images/uploads/technical_papers/PTI-32.pdf (accessed 11 October 2013).
ShortSea XML (n.d.), available at http://www.shortseaxml.org (accessed 28 August 2013).
SINA (2009), *Impacts Assessment of ITS Applications – Evaluation of the Dynamic Speed Control System on the Mestre Beltway*, Politecnico di Milano.
Šitavancová, Zuzana and Martin Hájek (2010), 'Thematic Research Summary: Intelligent Transport Systems', available at http://www.transport-research.info/Upload/Documents/201008/20100805_152456_33647_TRS_IntelligentTransportSystems.pdf (accessed 20 January 2014).
SMILE project measure report (2009), 'Bus priority system', available at http://www.its-toolkit.eu/2decide/toolkit/files/ITS_TOOLKIT_371_1318514646.pdf (accessed 3 October 2013).
Sundvor, Mariann (2007), *Transportbruker og Logistikk 2007*, available at http://www.google.gr/url?sa=t&rct=j&q=&esrc=s&source=web&cd=1&cad=rja&ved=0CDAQFjAA&url=http%3A%2F%2Fwww.konferanse.info%2Fmodules.php%3Fop%3Dmodload%26name%3DPagEd%26file%3Dindex%26p_deliver%3Dmedia%26m_id%3D74&ei=ZK1bUtWFAcLU4wT1jYGgCg&usg=AFQjCNGpdN_3m1QcQl52lCfu-8qqvkd3KA&sig2=ipTn0KMLHVZc_wH2myaVHA&bvm=bv.53899372,d.bGE (accessed 28 August 2013).

Sussman, J.M. (2005), *Perspectives on Intelligent Transportation Systems*, New York: Springer.

The Intelligent Container (n.d), 'The Intelligent Container: Linked intelligent objects in logistics', available at http://www.intelligentcontainer.com/ (accessed 11 September 2013).

Transport for London (TfL) (2007), *Central London Congestion Charging, Impacts Monitoring, Fifth Annual Report, July 2007*, London: Transport for London.

Transport for London (TfL) (2012), *Public and stakeholder consultation on a Variation Order to modify the Congestion Charging scheme*, London: Transport for London.

Tyrinopoulos, Yannis and Maria Chatziathanasiou (2010), *Freight Transport Innovations List*, Deliverable D2.2 of the EU-funded project ENABLE.

Tyrinopoulos, Yannis and Maria Chatziathanasiou (2011), *Roadmaps enabling the European know-how transfer to Brazil and Argentina*, Deliverable D3.1 of the EU-funded project ENABLE.

United Nations Economic Commission for Europe (UNECE) (2012), *Intelligent Transport Systems for sustainable mobility*, Transport Division.

United Nations Economic Commission for Europe (UNECE) (n.d.), 'Intelligent Transport Systems', available at http://www.unece.org/trans/theme_its.html (accessed 11 October 2013).

Van der Veen, Feiko and Henk Taale (2011), *Evaluation Field Trials with Dynamic Speed Limits*, evaluation report of the EU-funded project EasyWay.

Van der Veen, Feiko and Henk Taale (2013), *Urban Travel Times on VMSs around Eindhoven*, evaluation report of the EU-funded project EasyWay.

Virtanen, Niina, Anna Schirokoff, Juha Luoma and Risto Kulmala (2006), *Impacts of an automatic emergency call system on accident consequences*, Ministry of Transport and Communications Finland.

Watson, George, Steven Revill and Peter Cullen (2008), *Traffic Scotland Web Information Services*, EasyWay Secretariat and the Evaluation Expert Group.

Werner, Thomas (2012), *Public Transit Prioritization in Munich*, available at https://docs.google.com/viewer?a=v&q=cache:NdnDe_0lIl4J:sutip.mine.nu/TUMasia_its_singapore/Singapore2012_Werner.pdf+&hl=el&gl=gr&pid=bl&srcid=ADGEESha2ZXskC5Gfs2B5Y-NH7-Il-AiAbeqHzykH9ecp179qeDQUnLMT1HMKgqdPy2G6sWciTlwVbS1-2jry59ZCeXC-OM4yVQ95sODwUD3ytC6UTnuicvvBIY8uiDtEZm3iaz91ful&sig=AHIEtbQ0_7wRSWqQjzLvNqvrRNj808S6bA (accessed 2 October 2013).

Wiesenthal, T., G. Leduc, P. Cazzola, W. Schade and J. Köhler (2011), *Mapping innovation in the European transport sector*, European Commission, Joint Research Centre, Luxembourg: Publications Office of the European Union.

Zubillaga, Fernando (2011), *Bilbao prototype*, Deliverable FL.2.3 of the EU-funded project FREILOT.

APPENDIX: OVERVIEW OF COLLECTED GOOD PRACTICES ON ITS

ID	Name	Thematic area	Geographical area/ Country	Start date	Mode ID	Main technological elements
F1	Valencia Port Community System (PCS)	Information flow	SE/Spain	–	W	EDI, XML, web services
F2	Portbase	Information flow	WE/the Netherlands	2009	W	EDI, XML, Java, Oracle, operating system
F3	Ambrogio IT application	Information flow	SE/Italy	2007	MM	Software implementation
F4	Interporto Bologna	Information flow	SE/Italy	2008	MM	RFID, web services
F5	E-TRAIN and GOAL	Information flow	Europe	2006	Rl	GPS, XML, software implementation
F6	CESAR	Information flow	Europe	2004	MM	Internet portal technologies
F7	BRAVO train monitoring and Customer Information System (CIS)	Transport of goods	WE, SE/Brenner corridor (Munich–Verona via the Alps)	2005	MM	Software implementation, RFID
F8	ShortSea XML	Information flow	Europe	2008	W	UN CEFACT
F9	Southampton Vehicle Booking System (VBS)	Information flow	NE/UK	2005	W	Web browsers
F10	Ab Ovo RCS (Rail Cargo System)	Information flow	WE/the Netherlands, Belgium	–	Rl	Software implementation
F11	Aprixon ECHO	Transport of goods	WE/Germany	–	Rl	Software implementation
F12	RailOpt	Resource management	NE,WE/Switzerland, Luxembourg, Sweden	–	Rl	Java

ID	Name	Thematic area	Geographical area/ Country	Start date	Mode ID	Main technological elements
F13	Jerid RailMap	Information flow	Europe	–	Rl	Web services
F14	Intelligent Container	Transport of goods	Europe	–	MM	Web browsers
F15	HGV parking places availability on A10 Motorway	Professional drivers' information services	WE/France	–	Rd	Field sensors, software, VMSs
F16	ANPR system for heavy vehicle control	Traffic management	SE/Spain	2011	Rd	Cameras, control centre
F17	Telematics-Controlled Truck Parking	Professional drivers' information services	WE/Germany	2005	Rd	Detection devices, dynamic signs
F18	Geolocation of service vehicles	Safety and security	WE/France	–	Rd	GPS, control centre
F19	Traffic Wales Website – Freight Management	Information flow	NE/UK	2012	Rd	DATEX II
F20	Dynamic HGV Information	Safety and security	WE/Germany	–	Rd	Field sensors, information panels
F21	FREILOT Delivery Space Booking scheme	Transport of goods	SE/Spain	–	Rd	Loop detectors, Programmable Logic Controllers (PLSs)
F22	e-Freight	Information flow	Europe	2006	A	EDI, XML
F23	Electronic CSD (e-CSD)	Safety and security	Europe	–	A	EDI
F24	e-Air Waybill (e-AWB)	Information flow	Europe	2010	A	Standard RP1670
F25	Donau River Information Services (DoRIS)	Traffic management	WE/Austria	2006	W	Automatic Identification System
F26	PannonRIS	Traffic management	EU/Hungary	–	W	Automatic Identification System

P1	EU-SPIRIT	Traveller information service	NE, WE, EE/ Denmark, Germany, Luxembourg, Sweden, Poland, France		MM	Internet technologies
P2	AnachB.at	Traveller information service	WE/Austria	2009	Rd	Traffic telematics technologies, Graph Integration Platform (GIP)
P3	TMC, TMC plus	Safety and security	WE/Austria	2002	Rd	Radio broadcaster
P4	Traffic Control and Management System in AEBT – BRISA Group	Traveller information service	SE/Portugal	2011	Rd	CCTVs, VMSs, iBrisa/ ATLAS software
P5	Management of TCC and TICs	Safety and security	SE/Spain	–	Rd	CCTVs, weather stations, VMS, control centre
P6	Mobility Portal Rheinland-Pfalz	Traveller information service	WE/Germany	2004	Rd	Web services
P7	Remote traffic lights	Traffic management	WE/Belgium	2006	Rd	On-board devices, GPS
P8	Dynamic Speed Limits	Safety and security	WE/the Netherlands	2011	Rd	Loop detectors
P9	Reyers Tunnel	Traffic management	WE/Belgium	2007	Rd	CCTVs, VMSs, control centre
P10	Traffic Management System in highway sections	Safety and security	SE/Slovenia	2004	Rd	Detection systems, SCADA software, VMSs
P11	Dynamic Speed Control System	Safety and security	SE/Italy	2004	Rd	Detectors, VMSs, control centre
P12	Own Language Information System (OLIS)	Safety and security	NE/UK	–	Rd	Automatic Number Plate Recognition (ANPR) system, VMSs
P13	Traffic Scotland Web Information Services	Traveller information service	NE/UK	1999	Rd	Internet technologies

ID	Name	Thematic area	Geographical area/ Country	Start date	Mode ID	Main technological elements
P14	Multi-modal Traveller Information Service, Trafikken.dk/Hovedstaden	Traveller information service	NE/Denmark	2007	Rd	Internet technologies
P15	Urban Travel Times on VMSs	Traveller information service	WE/the Netherlands	2012	Rd	Loop detectors
P16	Meteotrans – European Weather Data Exchange	Information flow	WE/Germany, Luxembourg	2012	Rd	DATEX II, meteorological stations
P17	Highways Agency Weather Information Service (HAWIS)	Traveller information service	NE/UK	2012	Rd	Monitoring systems, web services
P18	Real Time Public Transport Information (RPTI)	Traveller information service	NE/Norway	2011	Rd	GPS
P19	Bus Priority System	Traffic management	NE/Norway	2011	Rd	GPS
P20	French–Spanish cross-border traffic management and information interoperability	Traffic management	WE, SE/France, Spain	2005	Rd	Monitoring systems, web services
P21	e-Call	Safety and security	Europe	2005	Rd	On-board devices, telecommunication, emergency centre
P22	Regional traffic information service	Traveller information service	NE/Finland	2009	Rd	Web services

P23	Automatic Incident Detection (AID) and fast alert	Safety and security	WE/France	2005	Rd	Cameras, control centre, VMSs
P24	Dynamic motorway lane management	Traffic management	WE/France	–	Rd	VMSs
P25	ERATO traffic management system	Traffic management	WE/France	2004	Rd	Optical fibres, VMSs
P26	CRITER	Traffic management	WE/France	–	Rd	Sensors, telecommunications, software implementation
P27	Totem	Safety and security	WE/France	2004	Rd	Processing unit, VMS
P28	Munich Tram and bus prioritization	Traffic management	WE/Germany	–	Rd	GPS
P29	eTraffic: Integration of traffic information and Google Maps	Traveller information services	SE/Spain	–	Rd	Web services
P30	Integrated system for optimizing the cashing and the costs in the Public Transport Company of Timisoara (RATT)	Traveller information service	EE/Romania	–	Rd	Automatic vehicle location system
P31	ITS applications on highway D8	Traffic management	EE/Czech Republic	2007	Rd	Detectors, VMSs, control centre
P32	Bilrejseplanen.dk	Traveller information services	NE/Denmark	2010	MM	Web technologies
P33	Traffic Information Agency Bavaria (VIB)	Traveller information services	WE/Germany	–	MM	Web technologies

ID	Name	Thematic area	Geographical area/ Country	Start date	Mode ID	Main technological elements
P34	Intelligent Urban Mobility Management System of Thessaloniki	Traveller information services	SE/Greece	2012	Rd	Loop and Bluetooth detectors, VMSs
P35	Fully automated free flow road toll stations in Tønsberg and Bergen, Norway	Electronic payment services	NE/Norway	2004	Rd	Electronic fee collection
P36	ARENA: A system for reporting road accidents	Information flow	SE/Spain	2005	Rd	Web services
P37	Bus Priority System	Traffic management	NE/Sweden	2007	Rd	GPS
P38	Congestion charging in London	Traffic management	NE/UK	2003	Rd	Automatic Number Plate Recognition (ANPR) system
P39	NADICS Journey Time Planner	Traveller information services	NE/UK	2003	Rd	Floating Car Data
P40	UTMC, UK	Traffic management	NE/UK	–	Rd	Transport network management technologies
P41	PADIS	Information flow	Europe	–	A	EDIFACT, XML schemas
P42	StB Matchmaker	Information flow	Europe	–	A	Web services
P43	Rotterdam Vessel Traffic Services (VTS)	Traffic management	WE/the Netherlands	–	W	Surveillance Systems
P44	Trainline's smartcard scheme	Electronic payment services	NE/UK	–	Rl	Mobile phone and web services

PART IV

Providing the framework of ICT for transport

12. Amalgamating ICT with sustainable transport – building on synergies and avoiding contradictions

Nikolas Thomopoulos, Moshe Givoni and Piet Rietveld

OUTLOOK

Academics have acknowledged the promising prospects of ICT for transport (Banister and Hickman, 2006) but have also stressed the fact that the death of distance is premature, pointing out that transport will continue to play a significant role in spatial economy (Rietveld and Vickerman, 2004). The latter urges stakeholders to take technological optimism bias into account along with other common sources of bias in transport appraisal (Mackie and Preston, 1998) when assessing such innovative schemes. It is apparent thus that ICT for transport can be linked with the sustainable mobility paradigm (Banister, 2008) since it presents some important advantages. Given the potential negative impacts mentioned in the previous chapters though, the view of ICT should not be solely focused on the technological dimension or be overly optimistic, which would be a narrow perspective. The problems and challenges faced by and within the transport sector are rather more social and environmental in nature, rather than merely technological. Thus a 'technological fix' is not really where resources should be invested, but it should be human capital foremost. The focus should predominantly be on social change, a change in which technology and ICT have a clear role to play (Givoni and Banister, 2013). ICT allows us to work while travelling, thus reducing the need to speed up travel which has environmental, cost and equity implications, but it can also help us speed up travel through various traffic management systems and this in turn can simply generate more demand for travel. Thus technology is a tool, not a solution, in progressing sustainable transport and mobility, and how it is put to use will determine its contribution, which is not guaranteed to be positive. Consequently, the aim of this book is to highlight the existing technological optimism bias whilst widening the discussion and debate, and at the same time exploring in a more holistic and systematic way this important subject, highlighting the opportunities along with the interrelated threats and risks. As such it aims at making a contribution in the development of appropriate theoretical frameworks to guide decision making, since the ICT of today will not be the ICT of tomorrow.

GUIDING PRINCIPLE

ICT has great potential and it is anticipated that it will be used increasingly within the transport domain. The implications of this use, namely the positive or negative impacts as well as the extent of these impacts on sustainable transport largely depend on the selected broader policy path besides the specific transport policy one. In every path selected, whether continuing to pursue economic growth, moving towards a more environmentally conscious future or a socially inclusive policy, ICT has an important role to play. The remaining question is how innovative technologies are being used.

The synchronous meaning of 'smart' in policy terms refers to the increasing use of ICT in a variety of ways to meet various objectives, ranging from social cohesion to economic growth and environmental sustainability. This is certainly the case in smart cities, for example, which feature at the heart of contemporaneous EU policy (DG Regio, 2011; Doranova et al., 2012; Foray et al., 2012) where transport is a focal point. Cities such as Amsterdam, Chicago, New York and Ottawa feature at the forefront of participatory design through Open Data competitions. Yet the contradictions that often arise when promoting these objectives do not disappear with the application of ICT, but are actually more likely to intensify the more successful the implementation of ICT is. This is often known as the direct rebound effect, which can lead to adverse outcomes of up to 30 per cent if not taken into account (Sorrell et al., 2009). An additional factor which needs to be considered is the magnitude of the informal economy, which accounts for a substantial proportion of GDP in developing countries, since the introduction of ICT often refers to the needs of those operating within the confines of the formal economy, neglecting those who do not have access despite their willingness to participate by using new technologies. Therefore, it is imperative to introduce a techno-social option which will make better use of Open Data, allowing wider adoption of ICT by all categories of travellers, hence increasing social inclusion (Greenfield, 2006).

In contrast to the ITU (2014) definition (Chapter 1, section 1.2), which does not target solutions, smart cities or smart transport systems should not be measured by the amount of ICT for transport applications or by the amount of electrons moving to maintain mobility needs, but should be measured solely by the extent to which ICT for transport is used to progress and meet transport policy objectives. ICT is the means, not the end, of transport policy, certainly in the case of sustainable transport policy, and the same goes for ephemeral terms like 'smart cities' which do not address directly fundamental issues such as wellbeing, freedom or security while on transit. The question is how ICT is applied, and this book serves the purpose of highlighting contemporary challenges, offering insight to foster collaboration between previously distant disciplines. Drawing on the preceding 11 chapters, the aim of this concluding chapter is to summarize findings and suggestions outlined to inform future policies founded on transdisciplinary synergies. Consequently, the conclusions elucidate the steps required to minimize identified threats and risks whilst maximizing synergies.

The guiding principle that should be drawn from this book and the one advocated for is the following: technology, and ICT in particular, is a core policy support tool in today's world. It has a key role to play and can definitely aid in the adoption of practically every transport policy, but the selection of appropriate policies should not be based or dependent on ICT. It can facilitate the increased use of public transport in urban areas (Pourbaix, 2012), for example through point-to-point public transport offered, for example, by Kutsuplus[1] in Helsinki, with positive environmental, economic and social impacts (NCE, 2014) aiding in moving away from the still widely used predict and provide model (Goulden et al., 2014). However, a cautious approach should be used to mitigate any threats or negative impacts which may include technological optimism bias fuelled by narrowly framed or overly optimistic proposed technological fixes and solutions.

SPECIFIC BOOK FINDINGS

Part I: Changing travel behaviour through ICT for transport shows that since travellers are not rational in their travel behaviour, the use of information will often not be rational either, resulting in a move away from, rather than closer to, beneficial use of information. Beneficial refers both to the individual traveller and to the group of users sharing the road at

---

1 More information available at: https://kutsuplus.fi.

a given time and location. 'Attitudes and affective stickiness' contribute to resisting acquisition of information and to changing travel behaviour when it has been acquired. However, information provided by ICT can certainly help to optimize traffic flow for example, and it can change travel behaviour in a desired way, but only when information is used through cooperation and social learning. It requires ICT systems to be designed and implemented with such cooperation and wider benefits in mind and not for the use of a single individual in competition with other individuals trying to gain benefits at their expense, as is the situation currently.

Furthermore, ICT for transport provides a unique opportunity to make public transport more inclusive and of better quality for certain social groups for example tourists, disabled and elderly travellers through the provision of real-time updates and alternatives. Employing custom designed incentives, has been proven – as for example in Chapter 4 through increased walking – to create lasting impact offering congestion relief and increased physical activity at low cost to address concurrent challenges such as obesity (UK Active, 2014). This can be achieved by incorporating ethical, social and legal issues through Privacy Impact Assessment or the involvement of Communities of Interest at the early design stages to improve penetration rates for diverse traveller categories as shown in Part III: Policies promoting ICT for transport: contemporary and future challenges. The latter will attain a future outcome which is not only environmentally sustainable but also socially acceptable.

Doing this by involving all stakeholders and glocal CoI while providing appropriate platforms for user-driven experimentation of ICT for transport, as discussed in Part II: ICT applications and sustainable transport: threats and opportunities, can convey positive outcomes in developed and developing countries, particularly when crowdsourcing and open data are used to generate and share transport information in innovative ways. The latter is particularly useful in bicycle planning as it offers a cost-effective solution with high accuracy. Benefits can also be encountered in fuel efficiency or average speed increase through ICT use and eco-driving in V2V communication, which can act as the predecessor of autonomous vehicles. Nevertheless, there are still challenges in reaching this stage since technological (for example transmission capacity equipment), safety and legal barriers need to be overcome first. Distributional issues through the unequal use of advanced ICT systems need also to be considered simultaneously to avoid negative impacts.

To date, most (64 per cent) of ICT for transport applications have focused on road transport, capitalizing on the business potential of surface transport modes as stated in Part III: Policies promoting ICT for transport: contemporary and future challenges. Nonetheless, this focus should also shift to include other transport modes, which would improve trust between stakeholders to share data and best practice. The latter can open up further business opportunities with wider benefits such as lower costs through competition and increased use, higher employment in this sector, as well as the inclusion of environmental and social objectives in existing business models. To achieve this though, it is essential to develop international standards, including open standards, and harmonized national policies through appropriate legislation and monitoring entities.

KEY CONCLUSIONS

Following the United Nations Climate Change Conference held in Durban in 2011, there has been some optimism within policy circles regarding the green economy, which has been substantiated during Rio+20 and continued after the 2014 UN meeting in New York. In tandem with the contemporary policy evolutions about low carbon development and the COP 21 in Paris, findings of this book are intertwined with the need to further enhance sustainable transport policies facilitated by ICT to support increased accessibility rather than increased mobility (Ferreira et al., 2012). Table 12.1 reflects on debates and academic discussions of the editors during the preparation of this book and shares recommendations which will contribute to shaping future research and policy agendas.

Table 12.1 Policy recommendations

Recommendations	Stakeholders
Common terminology. Involve diverse stakeholders in design phase.	Policy makers, Academics, Users, Corporations, Practitioners, International organizations.
Harmonized legislation.	Policy makers, International organizations.
Integrated technological platforms. Common standards.	Practitioners, Corporations, Users, International organizations.
User-friendly design and functionality.	Practitioners, Users, Policy makers.
Use of viable business models tested within various city or regional contexts.	Corporations, Policy makers, Corporations, Users.
Introduce appropriate incentives.	Users, Practitioners, Policy makers.
Share best practice regarding the provision of funds to develop and distribute ICT for transport.	Policy makers, Practitioners, Corporations.

A fundamental outcome is the need for all stakeholders to use common terminology across professions and academic disciplines. It is of equal importance to then integrate technological platforms and develop common standards for new ICT in an inclusive and efficient manner as has been done during the specification of, for example, Wi-Fi standards. Employing the knowledge and experience of both experts and users at an early stage, that is, during the design phase, should be a prerequisite. At the same time, by harmonizing legislation and introducing common standards internationally, such as ISO37120:2014 (ISO, 2014), it will become more feasible to form a dynamic and trustworthy Internet of Things (RAND, 2013), creating a universal environment for public authorities and private corporations to meet wider policy targets.

Despite not being the most important factor, design and functionality have also to be considered when developing ICT for transport (Kramers, 2014) since users have the capacity to turn certain infrastructure to an essential component of daily travel or to just another device with limited added value. Considering, for example, the needs of travellers who are visually or hearing impaired through certain infrastructure and smartphone applications (Matthews, 2014) constitutes a good example where groundbreaking progress could be achieved.

Furthermore, the global urbanization trend which anticipates that two-thirds of the global population will live in cities by 2050 (UN, 2014) offers a unique opportunity for cities to act as testbeds for ICT for transport applications through living labs. Based on the diverse context, geomorphology, quality of available infrastructure and user needs, this would allow key social and environmental threats to be identified whilst pointing out promising business models for diverse city types. Establishing sound data management procedures and exchanging know-how among urban and regional authorities within both developed and developing countries will accelerate this process and yield invaluable returns for users as well as the private and public sectors.

Such synergies will make better use of scarce resources and can lead to better dispersion of ICT for transport addressing key challenges worldwide. The use of incentives for users and stakeholders at different governance levels has to be harmonized in order to promote effective collaboration. For example, a dialogue on sustainable transport options may be strengthened by a new set of metrics and evaluation methods which allocate more weight to the achievement of co-benefits in the form of climate-friendly outcomes rather than merely focusing on reducing travel time on certain transport modes.

Hence, given the scale of investment needed and contemporary strains on budgets, national, regional and local governments have a key role to play in improving the enabling conditions for private investment in sustainable transport. Therefore, it would be sensible to improve the domestic framework for investment in urban mobility, and especially in urban transport, including various tools, instruments, policies and regulations, and not only through fiscal instruments, which are ineffective in certain contexts due to market distortions (NCE, 2014).

These conclusions are valuable for academics, practitioners and policy makers worldwide to recognize arising opportunities and appreciate upcoming threats when deploying applications amalgamating ICT with sustainable transport. Through evidence from around the globe it has been illustrated that a more participatory, adaptive, just, inherently resilient and inclusive approach (Greenfield, 2013) is required to build on synergies, avoid contradictions and make informed decisions in the rapidly evolving field of ICT for transport.

REFERENCES

Banister, D. (2008) The sustainable mobility paradigm, *Transport Policy*, 15(2), 73–80.

Banister, D. and Hickman, R. (2006) How to design a more sustainable and fairer built environment: transport and communications, *Intelligent Transport Systems, IEE Proceedings*, 153(4), 276–91.

DG Regio (2011) Regional policy for smart growth in Europe 2020, European Union Regional Policy, Brussels: European Commission DG Regio.

Doranova, A., Griniece, E., Miedzinski, M. and Reid, A. (2012) Connecting smart and sustainable growth through smart specialisation: A practical guide for ERDF managing authorities, Luxembourg: Publications Office of the European Union.

Ferreira, A., Beukers, E., and Brömmelstroet, M. (2012) Accessibility is gold, mobility is not: a proposal for the improvement of Dutch transport-related cost-benefit analysis, *Environment and Planning B*, 39, 683–97.

Foray, D., Goddard, J., Beldarrain, X., Landabaso, M., McCann, P., Morgan, P., Nauwelaers, C. and Ortega-Argiles, R. (2012) Guide to research and innovation strategies for smart specialisations (RIS3), Luxembourg: Publications Office of the European Union.

Givoni, M. and Banister, D. (2013) *Moving Towards Low Carbon Mobility*, Cheltenham, UK and Northampton, MA, USA: Edward Elgar Publishing.

Goulden, M., Ryley, T. and Dingwall, R. (2014) Beyond 'predict and provide': UK transport, the growth paradigm and climate change, *Transport Policy*, 32, 139–47.

Greenfield, A. (2006) *Everyware: The Dawning Age of Ubiquitous Computing*, Berkeley: New Riders.

Greenfield, A. (2013) *Against the Smart City*, New York: Do.

ISO (2014) Sustainable development of communities – Indicators for city services

and quality of life, ISO37120:2014(en). Accessed 24 June 2014 at: https://www.iso.org/obp/ui/#iso:std:iso:37120:ed-1:v1:en.
Kramers, A. (2014) Designing next generation multimodal traveler information systems to support sustainability-oriented decisions, *Environmental Modelling & Software*, 56, 83–93.
Mackie, P. and Preston, J. (1998) Twenty-one sources of error and bias in transport project appraisal, *Transport Policy*, 5(1), 1–7.
Matthews, B. (2014) Elderly and disabled travelers, in Garrett, M. (ed.) *Encyclopedia of Transportation: Social Science and Policy*, Thousand Oaks: SAGE, pp. 462–66.
NCE (2014) Better growth, better climate: The new climate economy report. Global report, Washington, DC: Global Commission on the Economy and Climate.
Pourbaix, J. (2012) Towards a smart future for cities, *Journeys*, May.
RAND (2013) Towards a dynamic and trustworthy Internet of Things, Research Brief RB-9742-EC, RAND Europe. Accessed 24 June 2014 at: http://www.rand.org/content/dam/rand/pubs/research_briefs/RB9700/RB9742/RAND_RB9742.pdf.
Rietveld, P. and Vickerman, R. (2004) Transport in regional science: The 'death of distance' is premature, *Papers in Regional Science*, 83(1), 229–48.
Sorrell, S., Dimitropoulos, J., and Sommerville, M. (2009) Empirical estimates of the direct rebound effect: A review, *Energy Policy*, 37(4), 1356–71.
UK Active (2014) Turning the tide of inactivity, London: UK Active.
UN (2014) World Urbanization Prospects, the 2014 revision, United Nations Department of Economic and Social Affairs, Population Division.

Index